Computer–Assisted
Reservoir Management

Copyright © 2000 by
PennWell Corporation
1421 South Sheridan Road/P.O. Box 1260
Tulsa, Oklahoma 74101

Cover & book design by Morgan Paulus

Library of Congress Cataloging-in-Publication Data

1   2   3   4   5   04   03   02   01   00

Computer–Assisted
Reservoir Management

Abdus Satter
Jim Baldwin
Rich Jespersen

table *of contents*

list *OF FIGURES*

list OF TABLES

preface

This book grew out of a course developed and presented at Texaco over the past few years. It was eventually presented as an SPE short course with the same title. Our goal is to provide the fundamental background for the concepts and to share examples of the computerized techniques that we believe are the key to optimizing the management of oil and gas reservoirs in today's competitive business environment.

The concepts discussed here are not new. Geoscientists and engineers have long dreamed of the ability to create mathematical models of their reservoirs with which they could try out various operating scenarios before actually implementing the most profitable plan. The tremendous improvements in computer performance have made it possible to enhance software to fulfill those dreams.

Previously in the domain of specialists using expensive mainframe computers, many valuable software tools are now readily available on the desktop of the practicing geoscientist and petroleum engineer. This is both a blessing and a curse. It allows the analysis of reservoir data to be done very rapidly and much more thoroughly than previously. However, at the same time, it places a greater responsibility on the practitioner to understand the principles behind a wide variety of reservoir management tools. It is all too easy to put some numbers into the "black box" and come up with results that are quite meaningless. Hopefully, the basics presented in this book will serve to inform the reader about the vast array of software tools available to assist him or her. We also hope to be able to stimulate a greater communication among colleagues of all disciplines to ensure that the data are thoroughly understood and software is used as intended. Teamwork is more important than ever if we are to manage our reservoirs in a way that will make our projects profitable and our companies successful.

Abdus Satter　　　*James O. Baldwin*　　　*Richard A. Jespersen*

acknowledgments
ACKNOWLEDGMENTS

The authors wish to acknowledge the support and permission of Texaco to publish this book. We also would like to thank our many industry colleagues, coworkers, and students, from whom we have learned about various aspects of reservoir management and the role of computer software in carrying out that task. In particular, our long association with Scientific Software-Intercomp (SSI, now part of Baker Hughes Inc.) and their integrated Petroleum WorkBench software has given us many examples to illustrate our ideas.

We especially thank Mr. Jack DeLage (formerly of SSI) for his contributions on geostatistical analysis in Chapter 8, Dr. Alain Gringarten of Imperial College, London (also formerly of SSI), for material from his course on well test analysis that we used in Chapter 9, and Dr. James Buchwalter and Dr. Raymond Calvert of Gemini Solutions for providing their Merlin and Apprentice reservoir simulation software and for their guidance in developing the examples in Chapter 19.

Special thanks and appreciation go to our wives, Yolanda Satter, Kathy Baldwin and Sandy Jespersen, and our families for their patience, understanding, and encouragement during the long period of planning and writing this book.

Lastly, the authors (two engineers and a geoscientist) worked together as an integrated team to produce this worthwile book, which brings out the role, use, and importance of the mulitdisciplinary, integrated computer software in managing the reservoirs.

Abdus Satter   *James O. Baldwin*   *Richard A. Jespersen*

foreword
FOREWORD

W. John Lee
Peterson Chair and Professor of Petroleum Engineering
Texas A&M University

Two trends in technology development have accelerated in importance in the petroleum industry during the last decade: Information Technology and Integrated Reservoir Management. This book provides a concise overview of the intersection of these technologies and their application in the industry's attempts to maximize the profitability of petroleum reservoirs.

One of my personal concerns as use of computer software has exploded in petroleum reservoir applications has been the tendency by some to regard software as a "black box" whose contents are very mysterious, virtually infallible, and universally applicable. The authors of this book share this concern, and deal with it directly by explaining in clear and simple language the physical and geological principles that underlie the types of software currently used in the reservoir management process. If the engineer or geoscientist will simply learn the underlying assumptions and applicable conditions of a given software package—erroneous conclusions derived from its use will diminish greatly.

Despite many years of emphasis on integrated teams developing and monitoring reservoir management plans, many of us still feel uncomfortable with our knowledge outside our limited areas of expertise. The authors understand this need, and have provided simple, clear explanations for the technical areas most often involved in reservoir management. I believe that the explanations will be useful to generalists and to specialists outside of their areas of expertise. The technical areas discussed include well log analysis, seismic data analysis, mapping and data visualization, geostatistical data analysis, pressure transient test analysis, and production performance analysis (including reservoir simulation).

Many of us have difficulty "keeping up" with available representative, quality software to solve problems that arise in reservoir management. Again, the authors have helped us. They provide illustrations of applications of typical software and, more importantly, provide listings of stand-alone and integrated software packages.

Finally, the authors have recognized the power of using examples as a teaching technique. The case studies in concluding chapters bring everything together: the problems, the reservoir management strategies for dealing with those problems, the software for implementing the strategies, and the results and decisions that form from this use of modern technology.

This book will prove to be of significant value to geoscientists and engineers in the petroleum industry.

ChapterONE
INTRODUCTION

OVERVIEW

Sound reservoir management practice[1] involves goal setting, planning, implementing, monitoring, evaluating, and revising unworkable plans. Success of a project requires the integration of people, technology, tools, data, and multi-disciplinary professionals working together as a well-coordinated team.

Integrated computer software plays a key role in providing reservoir performance analysis, which is needed to develop a management plan, as well as to monitor, evaluate, and operate the reservoir. It is also useful in day-to-day operational activities.

A major breakthrough in reservoir modeling has occurred with the advent of integrated geoscience (reservoir description) and engineering (reservoir production performance) software designed to manage reservoirs more effectively and efficiently. Several service, software, and consulting companies have developed and are marketing integrated software installed on a common platform. Users from different disciplines can work with the software cooperatively as a basketball team, rather than passing their data/results like batons in a relay race.

This book presents various techniques used in reservoir performance analysis. Since the results depend on the quality of the reservoir model, emphasis will be placed on how to build a more reliable reservoir model—involving geophysics, geology, petrophysics, and engineering. The role played by integrated computer software in developing a reservoir management plan, as well as in monitoring, evaluating, and operating the reservoir, will be discussed. The presentation will include integration of geoscience and engineering data and the use of integrated software for analyzing full-field performance, as well as single vertical or horizontal well production, coning well, and waterflood pattern performance.

SCOPE AND OBJECTIVE

This book is the third one in the series of the reservoir management books, which Satter and Thakur[1] started to publish several years ago. The first book, *Integrated Petroleum Reservoir Management*,[1] was followed by *Integrated Waterflood Asset Management*.[2] Certain basic information from the first book was used in the second book, and has also been used in the third book. In this way, the follow-up books can be treated as stand alone books.

A team of two engineers and a geoscientist wrote the current book in order to address both geoscience and engineering aspects of computer software used in reservoir studies. The authors possess more than 100 years of diversified domestic and international experience in reservoir studies, reservoir operations, and management.

This book is not written for just highly experienced engineers, but for geoscientists, field operation staffs, managers, government officials, and others who are involved with reservoir studies and management. It provides an overall knowledge of the various software in use for reservoir analyses. College students in petroleum engineering, geoscience, economics, and management can also benefit from this book.

Numerous geoscience and engineering computer software are used to provide reservoir performance analysis that is needed to develop a management plan, as well as to monitor, evaluate, and operate the reservoir. This book brings out the role, use, and importance of the computer software in managing the reservoirs. It is intended to help the user develop his own common sense approach how best to utilize computer software to manage reservoirs. There is *no* "one size fits all" approach as each reservoir is unique.

It will be the first book to illustrate how geoscience, engineering data, and computer software can be integrated to analyze reservoir performance under various scenarios for developing economically viable projects. Our intention is not to invent new wheels, but to package under one cover the knowledge and hands-on experience acquired through many years of our professional careers to demonstrate how the wheels run.

The book is not written to give the readers hands-on experience in running the software. It explains what the users need to know about the "black box" behind the software by providing the basic knowledge and understanding of the problems being solved. We provide examples of how different types of computer software are run, including the input data and output results. For example, it does not really teach how to carry out log interpretation, or how to carry out a pressure transient analysis. However, it does inform the reader about what types of software are available to do their particular job, and what can be expected.

In essence, it will present the following:

- Reservoir management concepts/methodology
- Techniques needed for reservoir performance analysis
- Examples to illustrate reservoir management using computer software

ORGANIZATION

The opening chapter presents an overview of the book, scope and objective, and organization.

Our stepwise thought process in organizing the book is given below:

- At the outset, chapter 2 on integrated reservoir management provides

the background necessary to grasp the use of computer software in managing the reservoirs.

- Chapter 3 deals with geoscience and engineering data needed to build the integrated reservoir model (chapter 4), which provides the foundation for reservoir performance analysis.
- Chapters 5, 6, 7, 8, and 9 on well log, seismic, mapping, geostatistics, and well test analyses provide the basic knowledge and applications of the computer software.
- Chapters 10, 11, 12, 13, and 14 deal with production performance analysis techniques, *i.e.*, volumetric, decline curve, material balance, and reservoir simulation.
- Chapter 15 provides an overview of the available stand alone and integrated computer software.
- Examples of full field studies around the reservoir cycle are given in chapter 16 for a newly discovered field development plan, chapter 17 for a mature field study, and chapter 18 for a waterflood project development.
- Mini-simulation can play an important role in everyday operations. Chapter 19 deals with examples of a single vertical or horizontal well, coning well, and waterflood pattern performance.

REFERENCES

1. Satter, A., and Thakur, G. C.: *Integrated Petroleum Reservoir Management: A Team Approach*, PennWell Books, Tulsa, Oklahoma (1994)
2. Thakur, G. C., and Satter, A.: *Integrated Waterflood Asset Management*, PennWell Books, Tulsa, Oklahoma (1998)

Chapter *Two*
RESERVOIR MANAGEMENT CONCEPTS AND PRACTICE

Integrated reservoir management has received significant attention in the industry in recent years through various technical articles and sessions, seminars, and even publication of a book.[1] The need to increase recovery from the vast amount of remaining oil- and gas-in-place worldwide requires better reservoir management practices to maintain a competitive edge. The principal goal of reservoir management is to maximize profits from a reservoir by optimizing recovery while minimizing capital investments and operating expenses. Economically successful operation of a reservoir throughout its entire life from discovery to abandonment requires:

- Integration: merging geoscience and engineering people, technology, tools, and data

- Synergy: multidisciplinary professionals working as a well-coordinated "basketball team" rather than as a "relay team"
- Support: company culture and organization removing barriers and fostering teamwork and integration

Satter and Thakur presented in their book[1] sound reservoir management concepts and methodology, technology needed for better reservoir management and examples to illustrate effective reservoir management practices. Satter, Wood, and Ortiz[2] presented a brief review of the conventional as well as the state-of-the-art concepts and approaches for integrated asset optimization and, more importantly, focus on the practice of improved asset management with real life examples.

This chapter discusses briefly the fundamentals of integrated reservoir management. It is provided because the readers need to know about the reservoir management goal, process and methodology so that they can have a better understanding of the role of computer software in managing the reservoirs.

RESERVOIR MANAGEMENT DEFINITION

Reservoir management can be defined as the process used to add value to the asset. It is no longer enough to just manage the reservoir itself. The focus is now on adding value to company assets by managing them in the context of the upstream, midstream, and downstream businesses. Integrated efforts and close working relationships between professionals knowledgeable in the areas of reservoir management, engineering, basic science, research and development, service, environment, land, legal, finance, and economics are essential.

HISTORY OF RESERVOIR MANAGEMENT

Reservoir management has advanced through various stages in the past 30 years. Before 1970, reservoir engineering was considered to be the most important technical aspect of reservoir management. During the 1970s and 1980s, the benefits of synergism between engineering and geology were recognized; the benefits of detailed reservoir description, utilizing geological, geophysical, and reservoir simulation concepts was realized.

RESERVOIR LIFE PROCESS

A reservoir's life begins with exploration, followed by discovery, delineation, development, production, and finally abandonment (Fig. 2-1). Multidisciplinary professionals, technologies, and tools are involved in the reservoir work.

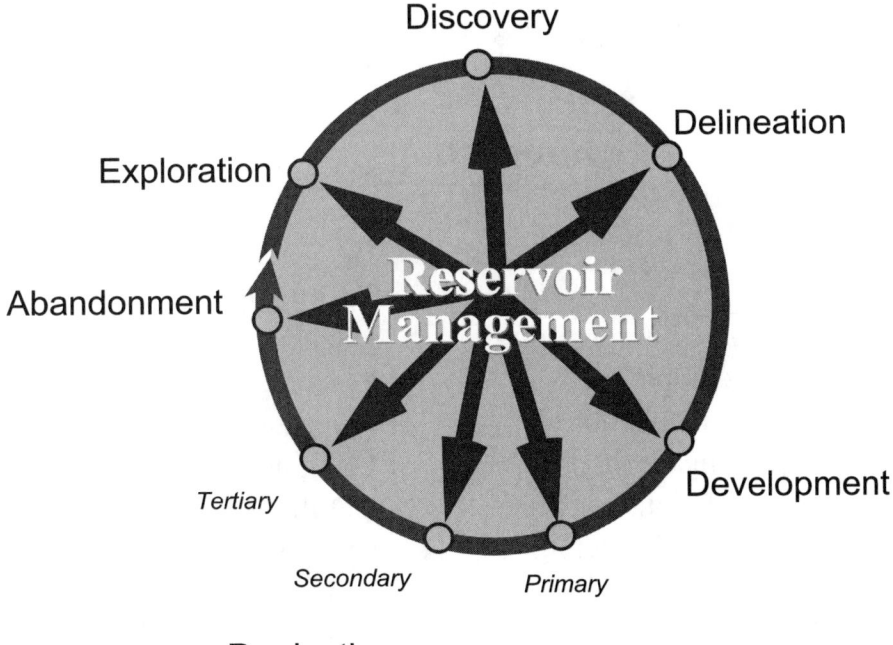

Fig. 2-1 • Reservior Life Process

INTEGRATION AND TEAMWORK

Successful operation throughout the life of the reservoir requires integration of geoscience and engineering, *i.e.*, people, technology, tools, and data (Fig. 2-2). It also requires synergy, *i.e.*, multi-disciplinary professionals working together as a team, rather than as individuals (Fig. 2-3).

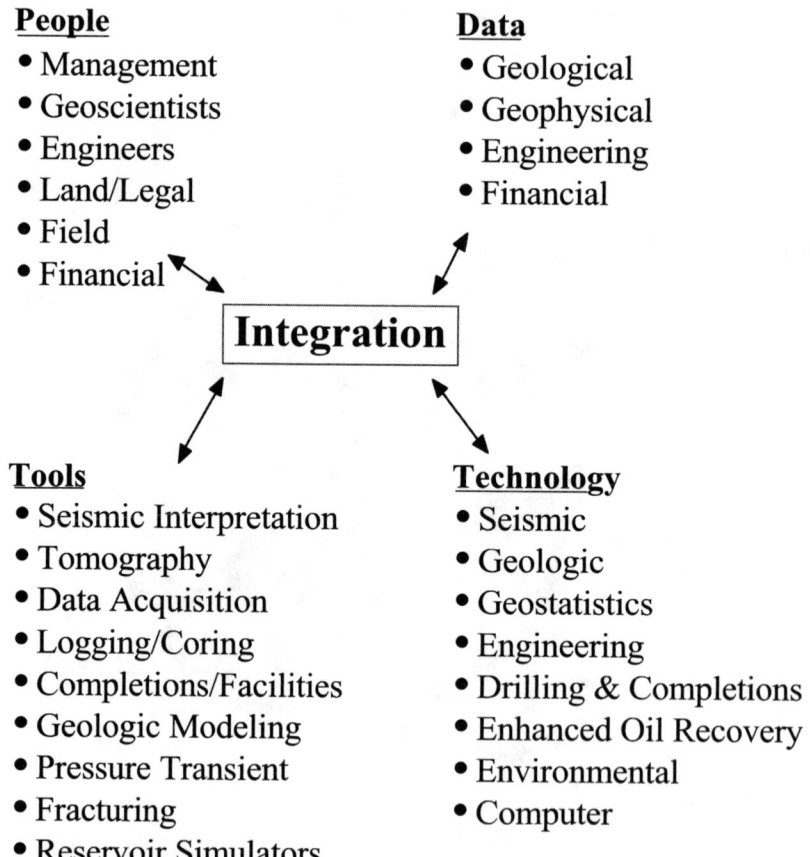

People
- Management
- Geoscientists
- Engineers
- Land/Legal
- Field
- Financial

Data
- Geological
- Geophysical
- Engineering
- Financial

Integration

Tools
- Seismic Interpretation
- Tomography
- Data Acquisition
- Logging/Coring
- Completions/Facilities
- Geologic Modeling
- Pressure Transient
- Fracturing
- Reservoir Simulators
- Enhanced Oil Recovery
- Computer Software & Hardware

Technology
- Seismic
- Geologic
- Geostatistics
- Engineering
- Drilling & Completions
- Enhanced Oil Recovery
- Environmental
- Computer

Fig. 2-2 • Integration of Geoscience and Engineering

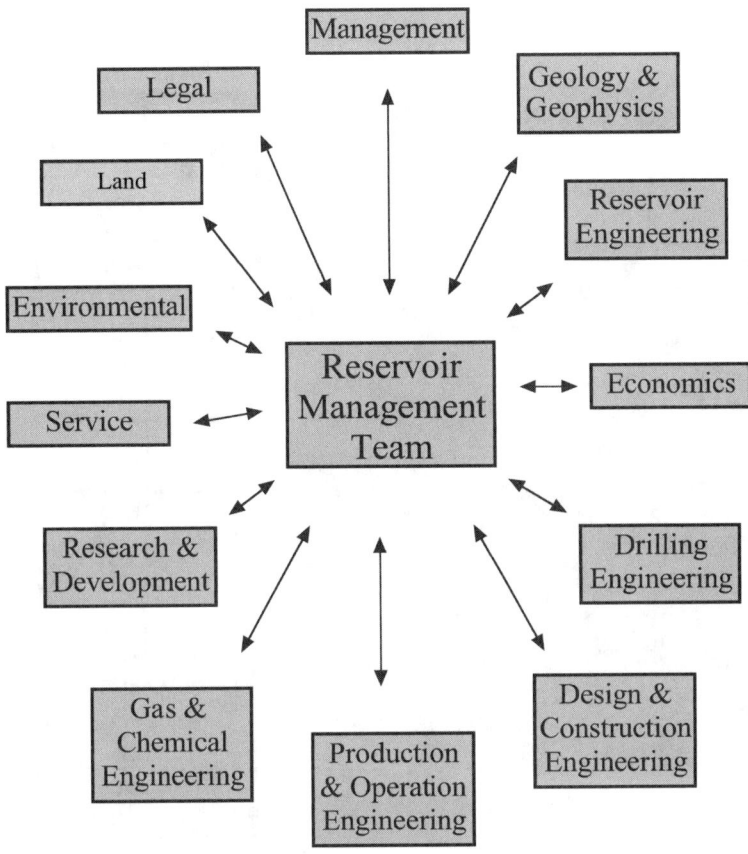

Fig. 2-3 • Multi-Disciplinary Reservoir Management Team

ORGANIZATION AND TEAMWORK

The organization and management of assets is also very critical. In the old system, various cross-disciplinary members of a team worked on an asset under their own functional management (Fig. 2-4). In the new multidisciplinary team approach, asset management team members from various functions work with a team leader whose responsibility is to coordinate all activ-

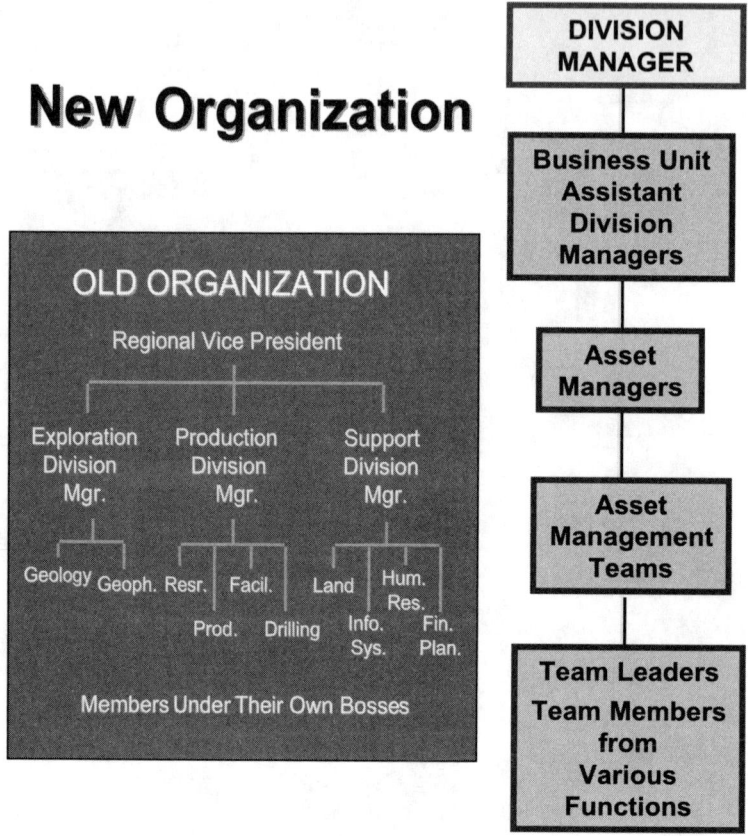

New Organization

OLD ORGANIZATION

Regional Vice President

Exploration Division Mgr.   Production Division Mgr.   Support Division Mgr.

Geology Geoph. Resr. Facil. Land Hum. Res.
                Prod. Drilling Info. Sys. Fin. Plan.

Members Under Their Own Bosses

DIVISION MANAGER

Business Unit Assistant Division Managers

Asset Managers

Asset Management Teams

Team Leaders
Team Members from Various Functions

Fig. 2-4 • Old and New Organizations

ities and report to the asset manager (Fig. 2-4). Administrative and project guidance is provided by the production manager or the asset manager. The asset management concept emphasizes focussing on a field(s) as an asset and all team members have the primary objective of maximizing the short- and long-term profitability of the asset.

Technology departments had to better align themselves closely with operating divisions of the company. Functional and organizational changes have been made within technology organizations in order to serve the busi-

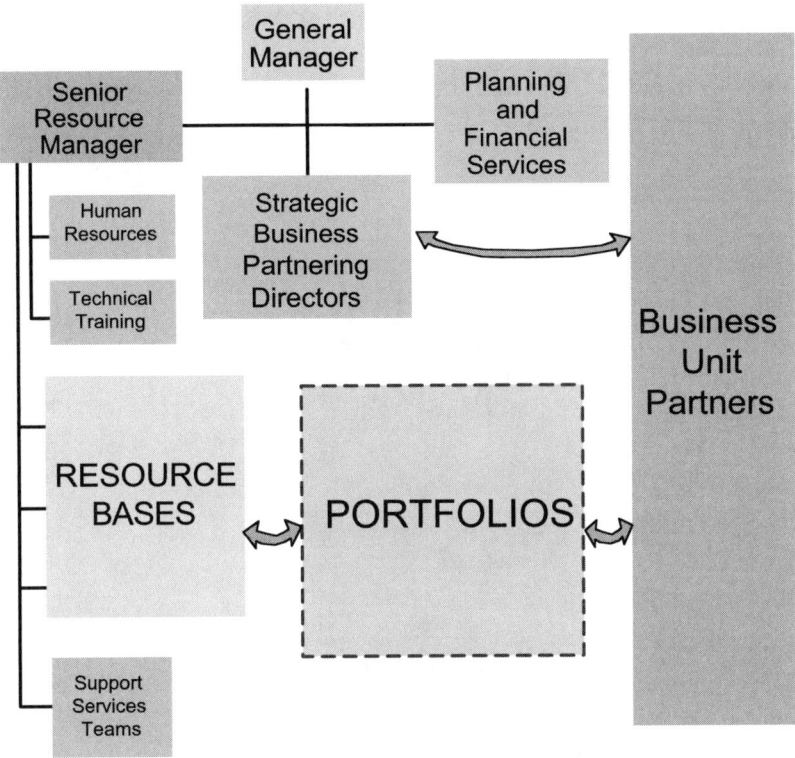

Fig. 2-5 • Texaco E & P Technology Organization

ness units better and add value to the company assets. Figure 2-5 shows an example of Texaco's E & P Technology Department organization, which is based on customer focused portfolios. The role of the portfolios in Texaco's organization is to design, develop, and deliver technology products and technical support.

RESERVOIR MANAGEMENT PROCESS

The modern reservoir management process consists of setting the goal, developing, implementing, and monitoring the plan, then evaluating the results (Fig. 2-6). No component of the process is independent of the others, therefore, integration of these components is essential for successful results. It is dynamic, ongoing, and no different than any other management process, such as financial, health, or business.

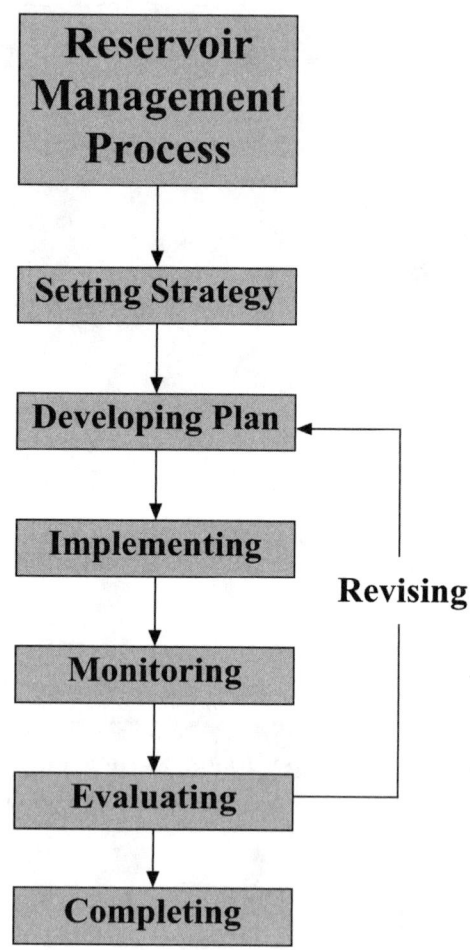

Fig. 2-6 • Reservoir Management Process

Setting the goal

The reservoir management goal can vary depending upon the company strategy being supported. Some examples include:

- Maximizing the economic value of an asset (Fig. 2-7)
- Maximizing employment in the national industry
- Conserving natural energy effectively

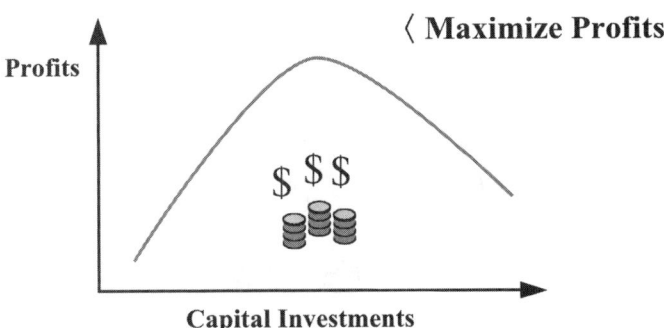

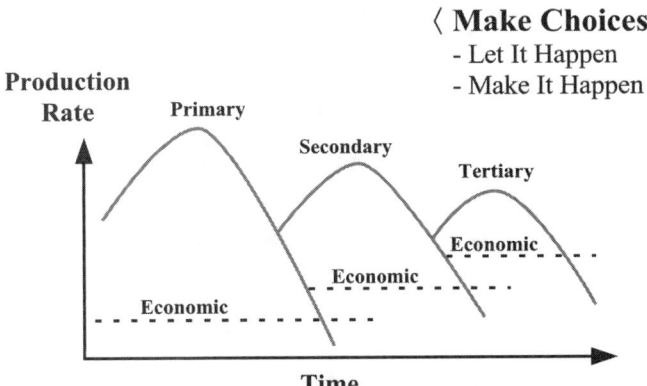

Fig. 2-7 • What is Reservoir Management?

In all of these cases, the conventional wisdom has been to optimally use all available resources, *i.e.*, human, technological, data, and financial, to maximize profits from a reservoir by optimizing recovery while minimizing capital investments and operating expenses.

In more recent years, the trend has been to set the goal and make it happen by whatever means possible, *e.g.*, new technologies, improved team work, joint ventures, and inter- and intra-company partnering and alliances.

Developing a plan

A comprehensive planning process will greatly improve the chances of successful reservoir management. It needs to be carefully worked out, involving many time-consuming development steps (Fig. 2-8).

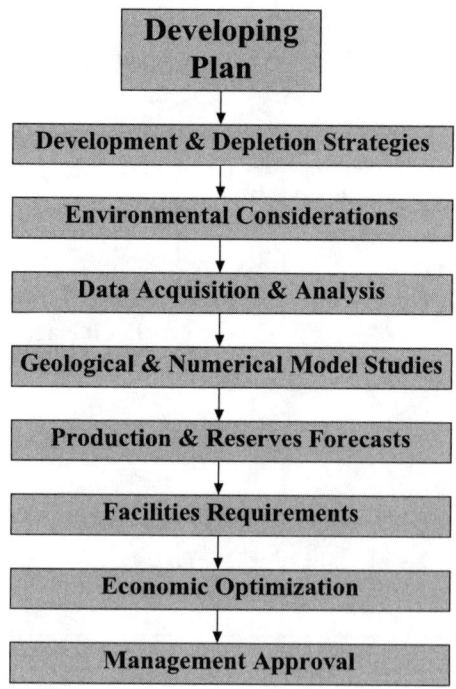

Fig. 2-8 • Developing Plan

The nature of the reservoir being managed is vitally important in setting its management strategy. Understanding the nature of the reservoir requires knowledge of the geology, rock and fluid properties, fluid flow and recovery mechanisms, drilling and well completion, and past production performance (Fig. 2-9). Reservoir knowledge is gained through an integrated data acquisition and analysis program to be discussed in chapter 3.

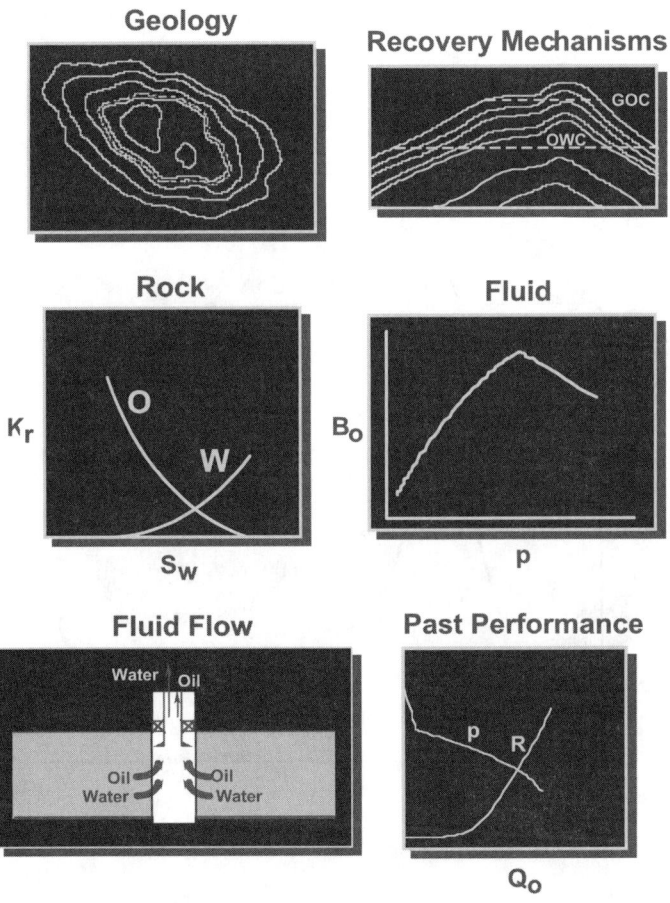

Fig. 2-9• Reservoir Knowledge

Traditionally, data of different types have been processed separately, leading to several different models—a geological model, a geophysical model, and a production/engineering model. The reservoir model is not just an engineering or geoscience model, rather it is an integrated model, prepared jointly by geoscientists and engineers, as will be discussed in chapter 4.

The economic viability of a petroleum recovery project is greatly influenced by the reservoir production performance under current and future operating conditions. Evaluation of the past and present performance and forecast of its future behavior are essential aspects of the reservoir management process (Fig. 2-10).

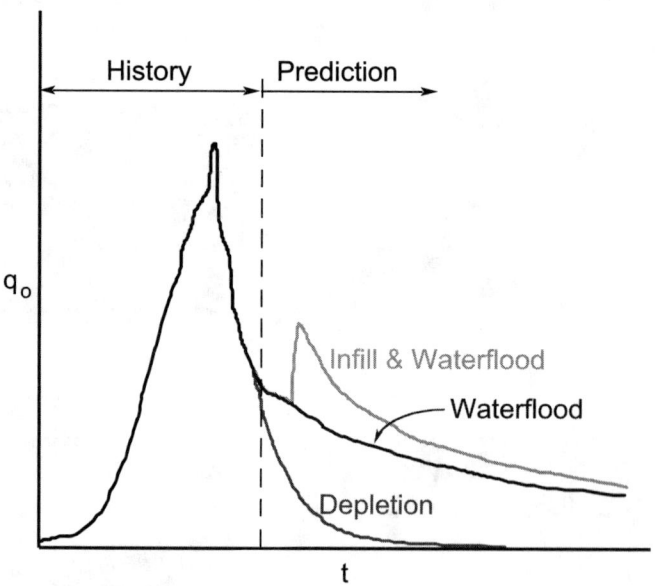

Fig. 2-10 • Production and Reserves Forecast

High-technology black oil, compositional, and enhanced oil recovery numerical simulators are now playing a very important role in reservoir management. As opposed to one reservoir life, the simulators can simulate many lives for the reservoir under different scenarios and thus provide a very powerful tool to optimize reservoir operation. Results of the forecasts are useful to monitor, evaluate, and operate the reservoir (Fig. 2-11).

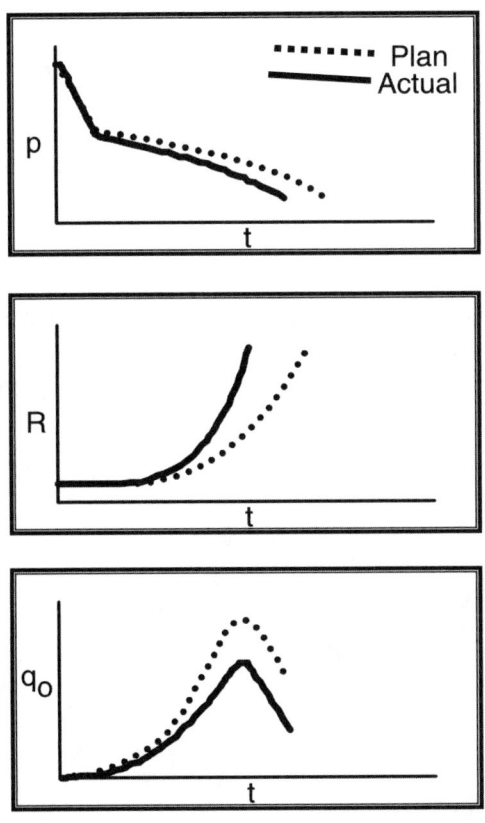

Fig. 2-11 • Performance Evaluation

Facilities that are the physical link to the reservoir include drilling, completion, pumping, injecting, processing, and storing. Proper design and maintenance of the facilities has a profound effect on profitability. The facilities must be capable of carrying out the reservoir management plan, but they cannot be wistfully designed.

Economic optimization is the ultimate goal selected for reservoir management. Chapter 18 presents an example illustrating the key steps involved in economic optimization.

Implementation

Once the goal and objective have been set and an integrated reservoir management plan has been developed, the next step is to implement the plan. Success in implementation can be improved by starting with a flexible plan of action, management support, and commitment of field personnel.

Surveillance and monitoring

Sound reservoir management requires constant monitoring and surveillance of the reservoir production, injection, and pressure data, in order to determine whether the actual performance is conforming to the plan (Fig. 2-11). A successful program requires coordinated efforts of the various functional groups working on the project.

Evaluation

How well is the reservoir management plan working? The answer lies in the careful evaluation of the project performance. The actual performance, e.g., reservoir pressure, gas-oil ratio, water-oil ratio, and production, needs to be compared routinely with the expected (Fig. 2-11). Any discrepancy should be resolved and adjustments made by utilizing integrated team efforts.

REFERENCES

1. Satter, A. and Thakur, G. C.: *Integrated Petroleum Reservoir Management: A Team Approach*, PennWell Books, Tulsa, Oklahoma (1994)
2. Satter, A., Wood, L., and Ortiz, R.: "*Asset Optimization Concepts and Practice*," JPT August, 1998

Chapter *THREE*

DATA ACQUISITION,
ANALYSIS, AND MANAGEMENT

The knowledge of the reservoir being managed is vitally important in setting reservoir management strategies. This chapter presents geoscience, engineering, and production/injection data needed to build the integrated reservoir model that will be discussed in chapter 4, providing the foundation for reservoir performance analysis. In addition, data acquisition, analysis, validation, storing, retrieval, and application will be presented in this chapter.

Throughout the life of a reservoir, from exploration to abandonment (Fig. 2-1), enormous amounts of data are collected. An efficient data management program, consisting of acquisition, analysis, validating, storing, and retrieving, can play a key role in reservoir management (Fig. 3-1). It requires planning, justification, prioritizing, and timing. An integrat-

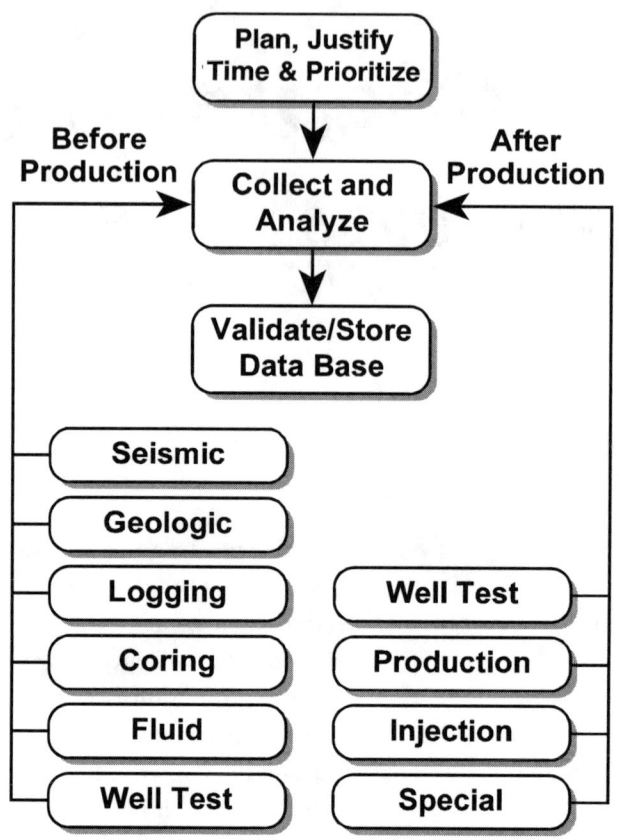

Fig. 3-1 • Data Acquisition and Analysis

ed approach involving all functions is necessary to lay down the foundation of reservoir management.

Satter and Thakur[1] provided a thorough discussion of this topic, which is summarized here.

DATA TYPE

The various types of data that are collected before and after production (Fig. 3-1) can be broadly classified as listed in Table 3-1.

- Seismic: 2-D, 3-D, and cross-well tomography

- Geological: Despositional environment, diagenesis, lithology, structure, faults, and fractures

- Rock: Logging - Open hole and cased hole
  Coring - Conventional, whole core, and side wall

- Fluid: Pressure-Volume-Temperature

- Well Test: Transient well pressure

- Production: Oil, gas, and water

- Injection: Gas and water

- Enhanced Data applicable to thermal, miscible, and
Oil Recovery: chemical floods.

Table 3-1 • Reservoir Data Classification

DATA ACQUISITION AND ANALYSIS

Multi-disciplinary groups, *i.e.*, geophysicists, geologists, petrophysicists, drilling, reservoir, production and facilities engineers, are involved in collecting various types of data throughout the life of the reservoirs. Land and legal professionals also contribute to the data collection process. Most of the data, except for the production and injection data, are collected during delineation and development of the fields.

An effective data acquisition and analysis program requires careful planning and well-coordinated team efforts of interdisciplinary geoscientists and engineers throughout the life of the reservoir. Justification, priority, timeliness, quality, and cost-effectiveness should be the guiding factors in data acquisition and analysis. It is essential to establish specification of what and how much data are to be gathered, and the procedure and frequency to be followed. It will be more effective to justify data collection to management, if the need for the data, costs, and benefits are clearly defined.

DATA VALIDATION

Field data are subject to many errors, *e.g.*, sampling, systematic, random, etc. Therefore, the collected data need to be carefully reviewed and checked for accuracy, as well as for consistency.

In order to assess validity, core and log analyses data should be carefully correlated and their frequency distributions made for identifying different geologic facies. The reservoir fluid properties can be validated by using equation of state (EOS) calculations and by empirical correlations. The reasonableness of geological maps should be established by using knowledge of the depositional environment. The presence of faults and flow discontinuities, as evidenced in a geological study, can be investigated and validated by pressure interference and pulse and tracer tests.

The reservoir performance should be closely monitored while collecting routine production and injection data, including reservoir pressures. If past production and pressure data are available, classical material balance techniques and reservoir modeling can be very useful to validate the volumetric original hydrocarbons-in-place and aquifer size and strength.

Laboratory rock properties, such as oil-water and gas-oil relative permeabilities, and fluid properties, such as pressure-volume-temperature (PVT) data, are not always available. Empirical correlations can be used to generate these data.

DATA STORING AND RETRIEVAL

The reconciled and validated data from the various sources need to be stored in a common computer database accessible to all interdisciplinary end users. As new geoscience and engineering data become available, the database requires updating. The stored data are used to carry out multipurpose reservoir management functions, including monitoring and evaluating the reservoir performance.

Storing and retrieval of data during the reservoir life cycle poses a major challenge in the petroleum industry today. The problems are:

- incompatibility of the software and data sets from the different disciplines
- lack of communication between databases

Many oil companies are staging an integrated approach to solving these problems. In late 1990, several major domestic and foreign oil companies formed Petrotechnical Open Software Corporation (POSC) to establish industry standards and a common set of rules for applications and data systems within the industry. POSC's technical objective is to provide a common set of specifications for computing systems, which will allow data to flow smoothly between products from different organizations and will allow users to move smoothly from one application to another.

DATA APPLICATION

Seismic data, especially 3-D, has gained tremendous importance in reservoir management in recent years as it provides valuable details of reservoir structure and faulting. Improvements in both recording and processing techniques, coupled with advancements in interpretation and visualization software, can reveal heterogeneities in complex reservoirs and even of the fluids within the rocks in some cases. The 3-D seismic information coupled with cross-well tomography also provides information on interwell heterogeneity.

Geological maps, such as gross and net pay thicknesses, porosity, permeability, saturation, structure, and cross-section, are prepared from seismic, core, and log analysis data. These maps, which also include faults, oil-water, gas-water, and gas-oil contacts, are used for reservoir delineation, reservoir characterization, well locations, and estimates of original oil- and gas-in-place.

The well log data, which provide the basic information needed for reservoir characterization, are used for mapping, perforations, estimates of original oil- and gas-in-place, and evaluation of reservoir perforation. Production logs can be used to identify remaining oil saturation in undeveloped zones in existing production and injection wells. Time-lapse logs in observation wells can detect saturation changes and fluid contact movement. Also, log-inject-logs can be useful for measuring residual oil saturation.

Unlike log analysis, core analysis gives direct measurement of the formation properties, and the core data are used for calibrating well log data. These data can have a major impact on the estimates of hydrocarbon-in-place, production rates, and ultimate recovery.

Core analysis is classified into conventional, whole-core, and sidewall analyses. The most commonly used conventional or plug analysis involves the use of a plug or a relatively small sample of the core to represent an interval of the formation to be tested. Whole core analysis involves the use of most of the core containing fractures, vugs, or erratic porosity development. Sidewall core analysis employs cores recovered by sidewall coring techniques.

The fluid properties are determined in the laboratories using equilibrium flash or differential liberation tests. The fluid samples can be either subsurface sample or a recombination of surface samples from separators and stock tanks. Fluid properties can be also estimated by using correlations.

Fluid data are used for volumetric estimates of reservoir oil and gas, reservoir type, *i.e.*, oil, gas or gas condensate, and reservoir performance analysis. Fluid properties are also needed for estimating reservoir performance, wellbore hydraulics, and flow line pressure losses.

The well test data are very useful for reservoir characterization and reservoir performance evaluation. Pressure build-up or falloff tests provide the best estimate of the effective permeability-thickness of the reservoir, in addition to reservoir pressure, stratification, and the presence of faults and fractures. Pressure interference and pulse tests provide reservoir continuity and barrier information. Multi-well tracer tests, used in waterflood and in enhanced oil recovery projects, give the preferred flow paths between the injectors and producers. Single well tracer tests are used to determine residual oil saturation in waterflood reservoirs. Repeat formation tests can measure pressures in stratified reservoirs, indicating varying degrees of depletion in the various zones.

Commonly used reservoir performance analysis and reserve estimation techniques that will be presented in chapters 10 through 14 are:

- volumetrics
- decline curves
- material balance
- mathemetical simulation

In addition to geological, geophysical, petrophysical, and reservoir engineering data, production and injection data are needed for reservoir performance evaluation.

REFERENCES

1. Satter, A. and Thakur, G. C.: *Integrated Petroleum Reservoir Management: A Team Approach*, PennWell Books, Tulsa, Oklahoma (1994)

Chapter *FOUR*

RESERVOIR MODEL

The economic viability of a petroleum recovery project is greatly influenced by the reservoir production performance under the current and future operating conditions. The accuracy of the results derived from the techniques used for analyzing reservoir performance and estimating reserves is dictated by the quality of the reservoir model used to make reservoir performance analysis. An integrated reservoir model, which is the subject of this chapter, requires a thorough knowledge of the geology, geophysics, rock and fluid properties, fluid flow and recovery mechanisms, drilling and well completions, and past production performance.[1] The basis for this knowledge is the data acquisition, analysis, and management, as discussed in the preceding chapter.

Role of Geoscience and Engineering

Traditionally, data of different types have been processed separately, leading to several different models—a geological model, a geophysical model, and a production/engineering model. The reservoir model is not just an engineering or a geoscience model—rather it is an integrated model, prepared jointly by geoscientists and engineers. The importance of geoscience and engineering data is discussed below. The many data sources required to adequately describe a reservoir can be seen in Table 4-1. The challenge is to integrate all this information into a model that reasonably describes reservoir performance.

Geoscience

Geoscientists probably play the most important role in developing a reservoir model. The distributions of the reservoir rock types and fluids determine the model geometry and model type for reservoir characterization.

The development and use of the reservoir model should be guided by both engineering and geological judgments. Geoscientists and engineers need feedback from each other throughout their work. For example, core analyses provide data to verify reservoir rock types, whereas well test analysis can confirm flow barriers and fractures recognized by the geoscientists. By discussing all the data as a team, each specialist can contribute the data he/she has available and can help other team members understand the significance of that data.

Three-dimensional seismic data can be used to assist in:

- defining the geometric framework
- qualitative and quantitative definition of rock and fluid properties
- flow surveillance

A 3-D seismic survey impacts the original development plan. With the drilling of development wells, the added information is used to refine the original interpretation. As time passes and the data builds, elements of the

DATA	SOURCE
STRUCTURE & ISOPACH MAPS	3D SEISMIC & WELL LOGS
POROSITY, PERMEABILITY, & FLUID SATURATIONS	WELL LOGS, CORES, & CORRELATIONS
FLUID CONTACTS & FORMATION TOPS	WELL LOGS
RESERVOIR PRESSURE & TEMPERATURE	WELL TESTS
PVT PROPERTIES	BOTTOM HOLE SAMPLES & CORRELATIONS
RELATIVE PERMEABILITIES	CORES & CORRELATIONS
PRODUCTION RATES & HISTORY	WELL TEST & ALLOCATION SUMMARY

Table 4-1 • Data Sources

3-D data that were initially ambiguous begin to make sense. The usefulness of a 3-D seismic survey lasts for the life of a reservoir.

Three-dimensional seismic surveys help identify reserves that may not be produced optimally. The analysis can save costs by minimizing dry holes and poor producers.

A 3-D survey shot during the evaluation phase is used to assist in the design of the development plan. With the development and production, data are constantly being evaluated to form the basis for locating production and injection wells, managing pressure maintenance, performing workovers, etc. These activities generate new information (logs, cores, DSTs, etc.), changing and/or revising previously generated maps, structures, stratigraphic models, etc.

Geostatistical modeling of reservoir heterogeneities is playing an important role in generating more accurate reservoir models. It provides a set of spatial data analysis tools as a probabilistic language to be shared by geologists, geophysicists, and reservoir engineers, as well as a vehicle for integrating various sources of uncertain information. Geostatistics is useful in modeling the spatial variability of reservoir properties and the correlation between related properties such as porosity and seismic velocity. A geostatistical model can then be used to interpolate a property whose average is critically important and to stochastically simulate for a property whose extremes are critically important.

Geostatistics enable geologists to put their valuable information in a format that can be used by reservoir engineers. In the absence of sufficient data in a particular reservoir, statistical characteristics from other more mature fields in similar geologic environments and/or outcrop studies are utilized. By capturing the critical heterogeneities in a quantitative form, one is able to create a more realistic geologic description.

ENGINEERING

After identifying the geologic model, additional engineering/production data are necessary for completion of the reservoir model. The engineering data include reservoir fluid and rock properties, well location and completion, well-test pressures, and pulse-test responses to determine well continuity and effective permeability. Material balance calculations can provide the original oil in place, and natural producing mechanisms—including gas cap size and aquifer size and strength. The use of injection/production profiles provides vertical fluid distribution.

INTEGRATION

Integration of geoscience and engineering data is required to produce the reservoir model, which can be used to simulate realistic reservoir performance (Fig. 4-1).

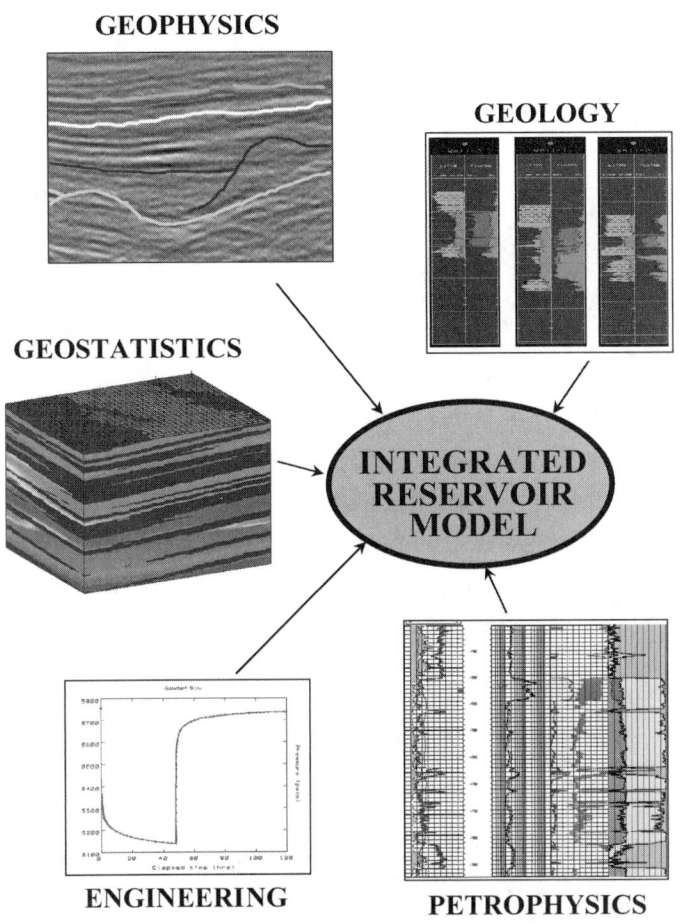

Fig. 4-1 • Integrated Reservoir Model

Development of a sound reservoir model plays a very important part in the reservoir management process because:

- It requires integration among geoscientists and engineers.
- It allows geoscientists' interpretations and assumptions to be compared to actual reservoir performance as documented by production history and pressure tests.

- It provides a means of understanding the current performance and to predict the future performance of a reservoir under various "what if" conditions so that better reservoir management decisions can be made.

In addition, the reservoir model should be developed jointly by geoscientists and engineers (Fig. 4-2) because:

- An interplay of effort results in better description of the reservoir and minimizes the uncertainties of a model. The geoscientists' data assist in engineering interpretations, whereas the engineering data sheds new light on geoscientists' assumptions.

- Assemble complete data set
- Share understanding of data
- Realize measurement uncertainties
- Refine model as new data are gained

Fig. 4-2 • Data Sharing and Validation

- The geoscientist-engineer team can correct contradictions as they arise, preventing costly errors later in the field's life.
- In a fragmented effort, *i.e.*, when engineers and geoscientists are not in communication with one another, each discipline may study only a fraction of the available data, thus, the quality of the reservoir management can suffer, adversely effecting drilling decisions and depletion plans throughout the life of the reservoir.
- Multi-disciplinary teams using the latest technology provide opportunities to tap unidentified reserves. For example, improved 3-D seismic data can aid in surveillance of production operations in mature projects and can identify presence or lack of continuity. between wells, and thus improves the description of the reservoir model
- One very important factor, which is often overlooked, is to discuss measurement uncertainties. Should a number be taken as hard fact, or is there a range of possible values around the given value? Utilizing reservoir models developed by multi-disciplinary teams can provide practical techniques of accurate field description to achieve optimal production.

A major breakthrough in reservoir modeling has occurred with the advent of integrated geoscience (reservoir description) and engineering (reservoir production performance) software designed to manage reservoirs more effectively and efficiently. Several service, software, and consulting companies are now developing and marketing integrated software installed in a common platform (chapter 15). These interactive and user-friendly software packages provide more realistic reservoir models. The users from different disciplines can work with the software cooperatively as a baseball team rather than passing their own results like batons in a relay race.

REFERENCES

1. Satter, A. and Thakur, G. C.: *Integrated Petroleum Reservoir Management: A Team Approach*, PennWell Books, Tulsa, Oklahoma (1994)

Chapter *FIVE*

WELL LOG ANALYSIS

INTRODUCTION

Measurements taken in the wellbore by wireline logging devices provide critical data for understanding and describing the reservoir. The properties displayed at measured depths on the log curves represent combined responses of the rocks and fluids in the formation, as well as tool geometry and the type of fluid in the borehole.

Typical measurements recorded on well logs include:

- spontaneous potential
- natural gamma radiation
- induced radiation
- resistivity
- acoustic velocity

- density
- caliper

But the ultimate analysis results needed for reservoir management are:

- producing zone depths
- zone thicknesses
- rock types
- porosities
- permeabilities
- fluid saturations

Most of the required information cannot be measured directly. It must be inferred from the log responses by making certain assumptions about the interaction of the logging device with the rocks and fluids.

Before this analysis can be done, the logs must be adjusted or "environmentally corrected" for the effects of the tool geometry and borehole fluids. Since any particular log may detect one reservoir condition and be insensitive to another, several types of logs are usually compared to determine the true nature of the reservoir.

Besides the digital curve values, additional information from the well is required for proper formation evaluation. Some of the additional data is found on the log header of the paper log. It includes:

- mud resistivity
- mud temperature
- mud density
- mud composition
- borehole temperature
- tool geometry

The comment portion of the log is used by the logger to report important information on the tool operation, such as:

- malfunctions
- hole conditions
- logging company name

It is necessary to know which logging company recorded the data, since the environmental corrections are different for each company's tools. The original log tape may contain all this information, but digitized log tapes usually do not. So the paper log headers and comments may be needed.

LOG TYPES

There is a large and growing number of logging tools available to aid in the determination of reservoir properties. Understanding the details of their operation and the subtleties of interpreting the measurements is the domain of the log analyst. Below we list a few of the basic logs and group them into categories according to the results they provide.

Caliper log

Measurement of borehole diameter by the caliper log reveals deviations from actual bit size. Increases in diameter indicate washouts of unconsolidated beds. Surprisingly, the borehole can also become smaller than the bit size. This results from the buildup of a mud cake in zones that are permeable enough to take the water from the drilling mud. The caliper is also used to indicate where pad type tools may give inaccurate readings due to either rapid change in hole size, or hole size too large for pad contact with the formation.

Other logs are broadly categorized into three types: lithology logs, resistivity logs, and porosity logs.

Lithology logs

Spontaneous potential (SP). This tool measures electrical currents that occur naturally when fluids of different salinities are in contact. Permeable zones are invaded by filtrates from the drilling mud. Since the salinity of the reservoir fluid generally is different than that of the drilling mud, SP currents are generated by the interaction of the two fluids. This tool is normally used with fresh water-based drilling muds.

The generated potential (in millivolts) is given by:

$$SP = -(60 + .133 \, T) \log (R_{mf} / R_w)$$

where:

T = temperature, °F

R_{mf} = mud filtrate resistivity, ohmmeter

R_w = water resistivity, ohmmeter

If all the other parameters are known, this equation can be solved for R_w. But relative values, rather than absolute SP readings, are used when interpreting the log curve. Impermeable shale zones can be used to set a "100 % shale base line" from which one can measure the relative magnitude in SP variations. The SP deflection is a qualitative indicator of permeability, but it is reduced by shale or hydrocarbon content.

Gamma ray (GR). Natural radiation of the formation (usually due to clay or shale content) is recorded by the gamma ray tool. It can be used when SP is not applicable and is therefore useful in detecting and evaluating deposits of radioactive minerals. In sedimentary formations, the gamma ray log, which has an average depth of penetration of one foot, reflects the overall shale content of the formation. This is due to the fact that the radioactive deposits usually concentrate in clays and shales. Clean formations generally have a very low level of radioactivity. This log is frequently used as a substitute for the SP in cased holes where the SP in unavailable or in open holes where the SP is unsatisfactory.

The gamma ray log can be recorded in cased holes. That makes it quite useful in completion and workover operations. It is often used for correlations and, when combined with a casing collar log, allows for accurate positioning of perforating guns. The gamma ray log is also often used in conjunction with radioactive tracer operations.

Resistivity logs

Much information about the reservoir can be deduced from measurements of electrical resistivity. The classic equation of Archie gives the water saturation, S_w, as a function of porosity, ϕ, water resistivity, R_w, and true formation resistivity, R_t:

$$S_w = \left(\frac{aR_w}{\phi^m R_t} \right)^{1/n}$$

Here, a, m, and n are experimentally determined constants. The parameters vary, but are often approximately $a = 1$, $m = 2$, and $n = 2$. This relationship works well in clean formations, but not as well in shaly formations or formations in which the connate water is fresh. Resistivity is used to determine water saturation (since only water in the rocks is usually conductive). The conductivity is proportional to the NaCl content of the water, to temperature and to water volume (porosity). Sometimes resistivity is used as the base log to pick formation tops and thicknesses and to correlate with other wells. The formation resistivity factor is defined as $F = R_o / R_w$, where R_o is the resistivity of a formation 100% saturated with water of resistivity R_w.

Electric log. This classic log consists of an SP curve and a combination of resistivity curves designated normal or lateral.

Induction-electric log. Formation conductivity is measured by this log, which is a combination of SP or GR, 18" normal, and induction curves. It works best for medium to high porosity.

Dual induction log. SP and/or GR and three resistivity curves (three depths of investigation) combine to make the dual induction log. Resistivity curve separation indicates invasion of drilling fluid into the formation. The difference is reduced, however, by the presence of hydrocarbons. The short device measures the flushed zone, R_{xo}, while the medium induction curve measures both flushed and invaded (R_i) zones, and the deep induction measures primarily the uncontaminated true reservoir resistivity (R_t). The ratios of shallow/deep and medium/deep yield d_i (diameter of invaded zone), R_{xo}, and R_t.

Guard log (laterolog). Guard electrodes focus the formation current into a thin disk in this log, which is used in conductive muds, thin beds, and high resistivity formations.

Porosity logs

Acoustic velocity log (sonic log). The difference in travel time, Δt, from a sound source at one end of the tool to two receivers at different distances along the tool is measured. This difference represents the travel time of sound through the portion of the reservoir between the two receivers. Both the rock matrix and the fluid in the pores influence the result, resulting in an estimate of the porosity, ϕ:

$$\phi = \frac{\Delta t_{\log} - \Delta t_{matrix}}{\Delta t_{fluid} - \Delta t_{matrix}}$$

This log is good for primary porosity, but does not indicate all secondary porosity. It yields high porosity in shaly zones.

Density log. A similar-looking estimate for porosity is obtained from the measurement of the electron density of the formation using a chemical source of gamma radiation and two gamma detectors:

$$\phi = \frac{\rho_{matrix} - \rho_{bulk}}{\rho_{matrix} - \rho_{fluid}}$$

where ρ represents density. This log works in air-drilled holes or with any type of fluid. Its penetration is shallow, so ρ_{fluid} is usually taken to be drilling fluid density (1.0 for fresh mud, 1.1 for salt-base mud). The density log results in pessimistic S_w in the presence of shale (unless corrected), but the porosity is not much affected by shale. On the other hand, the density log yields high porosity in the presence of gas.

Neutron log. Induced radiation is measured by bombarding the formation with fast moving neutrons. As these neutrons collide with hydrogen nuclei in the formation fluids, they lose energy and eventually are captured, causing the emission of a gamma ray. The number of gamma rays is proportional to the amount of hydrogen in the formation. Since the concentration of hydrogen in oil and water is similar, the result is an estimate of fluid-filled porosity. Shales indicate high neutron porosity due to bound water (can be used as a shale indicator). Gas zones indicate low neutron porosity.

Combination porosity logs. Comparing the measurements from two or more porosity tools can resolve the uncertainties presented by the individual devices in differentiating liquids from gas, identifying lithology, and determining shale volume. Both acoustic velocity and density have a weak response to gas, while neutrons indicate a much lower porosity for gas-filled rock than for fluid-filled rock. Density, acoustic velocity, and neutrons all respond differently to lithology. The ratios of density-derived porosity to acoustic velocity or neutron porosity yield the shale volume.

ANALYSIS PROCEDURE

The steps involved in performing the log analysis include data entry, data preparation, the analysis itself, and verification of the results, as detailed below:

1. Data entry
   - loading log curves
   - entering other logging information
2. Log data preparation
   - quality control, trace edit, and depth shift
   - borehole environmental corrections
3. Well log analysis
   - parameter selection
   - computation of reservoir properties
4. Verification of results with core, drill cutting, and other wells

The procedure is essentially the same (except for data loading) whether done with paper logs or by computer. The following example illustrates the procedure when an integrated software package is used.

EXAMPLE ANALYSIS WITH INTEGRATED SOFTWARE

Some integrated applications include not only the facilities needed to analyze well-log data for single wells, but also additional capabilities for analyzing entire reservoirs. For example, with Baker-Hughes/SSI's Petroleum WorkBench, the reservoir description module can perform all the following functions:

- Data entry

- Log data preparation
    - trace editing and depth shifting
    - environmental corrections
- Cross sections
- Integration of core analysis
- Well log analysis
    - parameter selections
    - saturation equation selection
- Summations
- Mapping and contouring
- Volumetrics

Note the similarity with the analysis procedure outlined above. A few key steps in the computer analysis are illustrated below:

Data entry

The log traces are imported as files or digitized from paper. Then supplementary information about the logging run, including mud properties and tool types are keyed into forms as shown in Figure 5-1.

Log data preparation

The raw logs are examined (as in Fig. 5-2) and corrections are made interactively to any bad data segments. Curves may also be shifted to align them properly with each other.

Well log analysis

Data contained in the logs can be graphically displayed in various ways (Fig. 5-3) to aid the determination of parameters and cutoffs for the analysis. Then the analysis is run, converting the recorded data into the desired logs of rock and fluid properties, as shown in Figure 5-4.

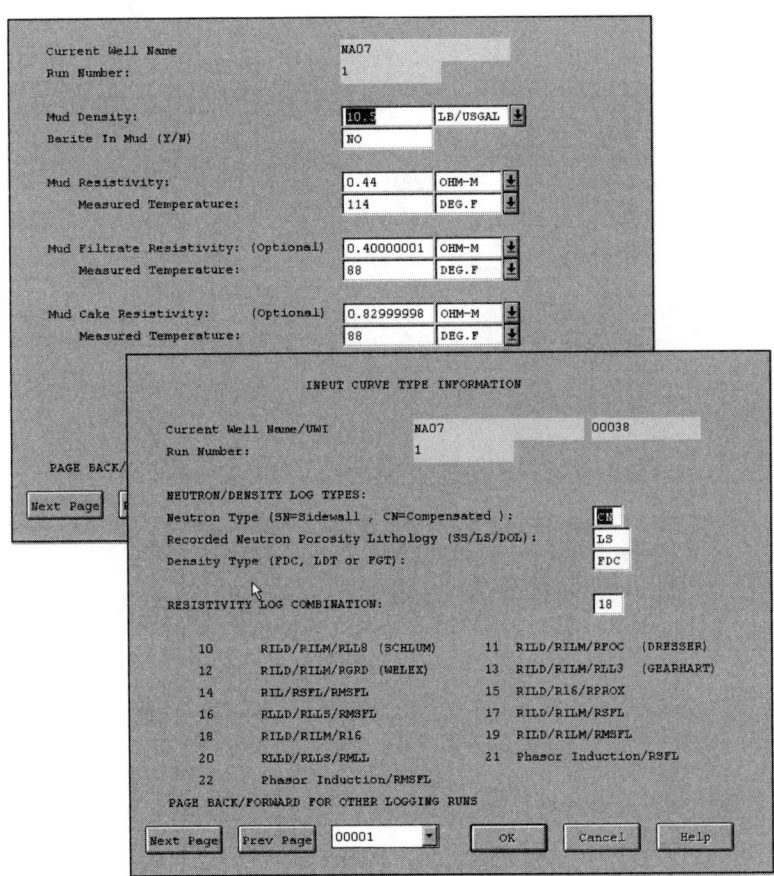

Fig. 5-1 • Entry of Logging Information

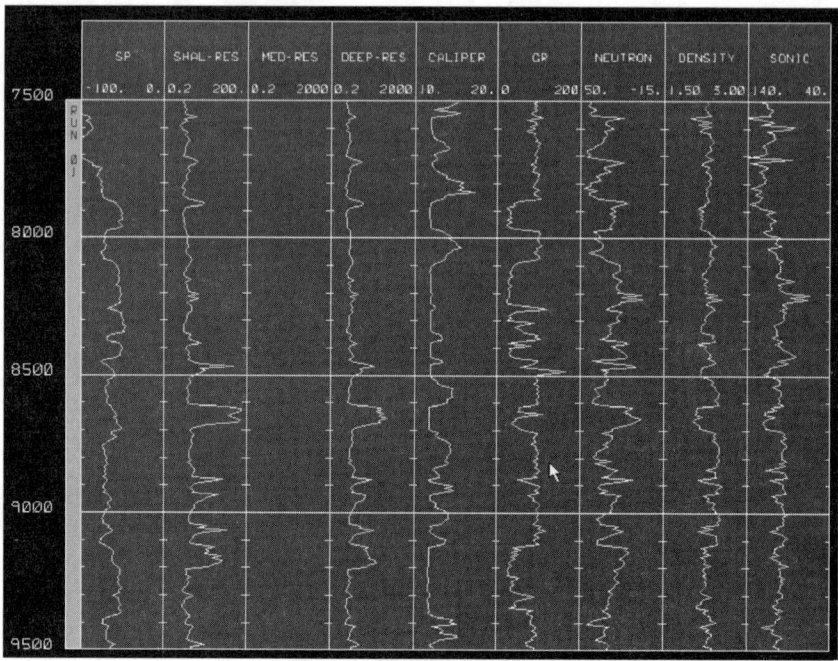

Fig. 5-2 • Raw Logs Ready for Analysis

Cross sections

Cross sections of the raw logs can be used to correlate geologic tops from well to well and isolate the zone to be analyzed. After the analysis, cross sections of results logs can be displayed as in Figure 5-5.

Summations

Geologic layers are defined on the cross sections. Then the summation process determines the representative value of each log property in each layer for each well. These results are tabulated for one well in Figure 5-6. The results for all wells are also automatically displayed on the associated prop-

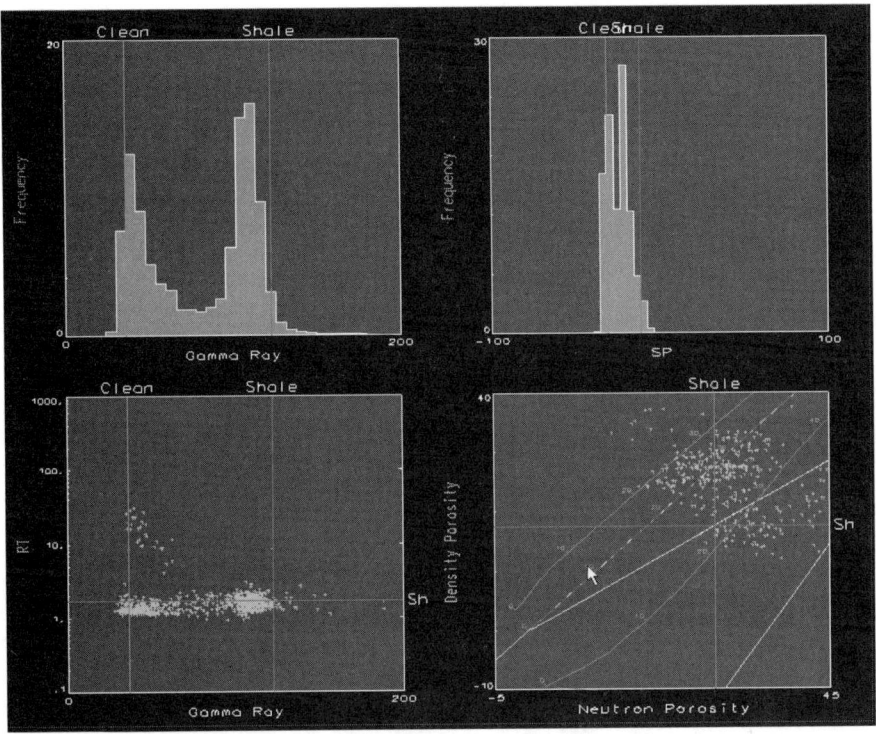

Fig. 5-3 • Analysis Plots for Lithology Parameters

erty maps of each property in each layer, ready for contouring. An example of one such property map is shown in Figure 5-7, which can be used to generate contour maps.

The beauty of this computer analysis is that all the steps are performed in one program, minimizing the often time-consuming task of loading and exporting data from one program to another.

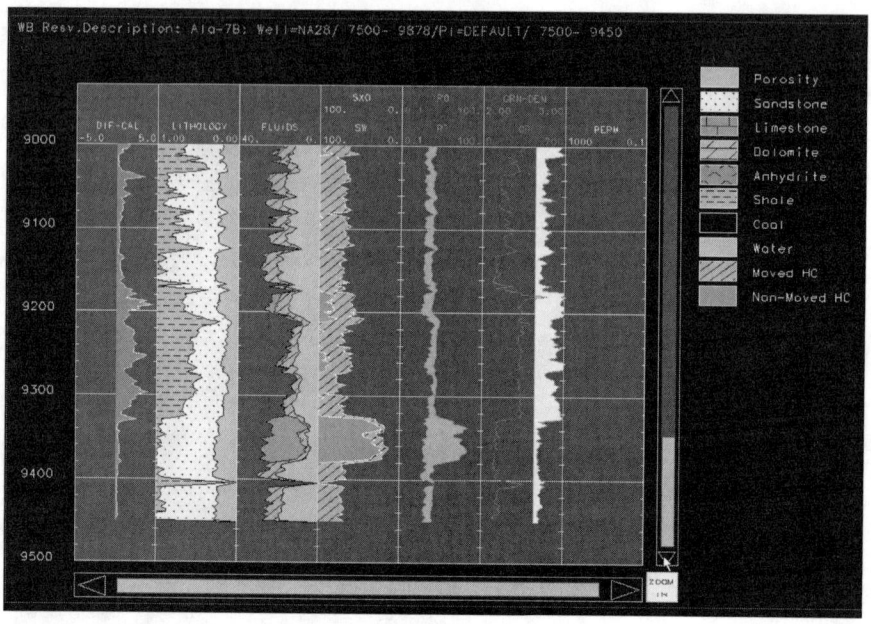

Fig. 5-4 • Results of Log Analysis

REFERENCES

Some good sources of additional information on log analysis are:

- *Introduction to Wireline Log Analysis*, Western Atlas
- *Log Interpretation Principles and Applications*, Schlumberger
- *Open Hole Analysis and Formation Evaluation*, Texaco

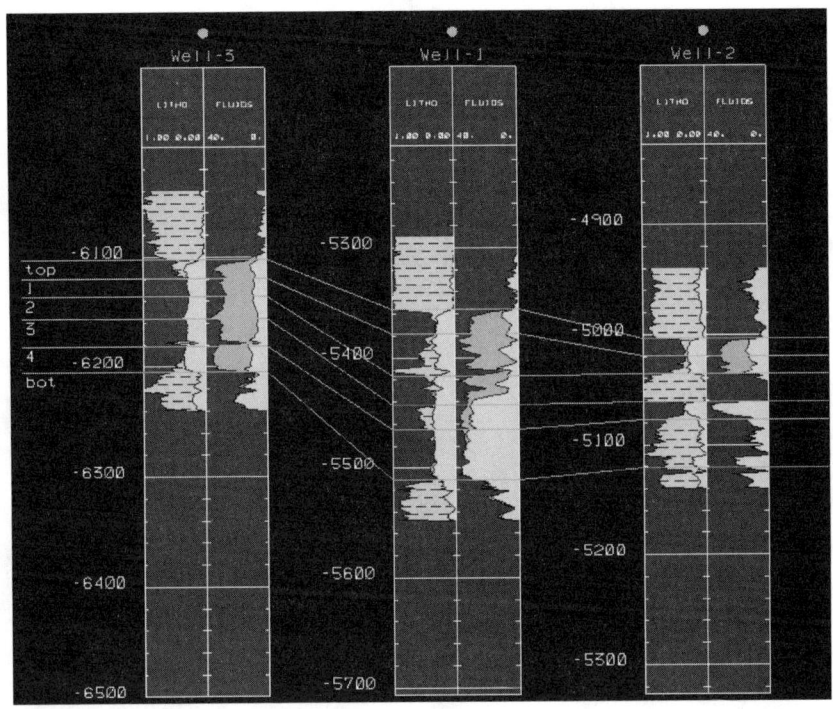

Fig. 5-5 • Log Cross-Section

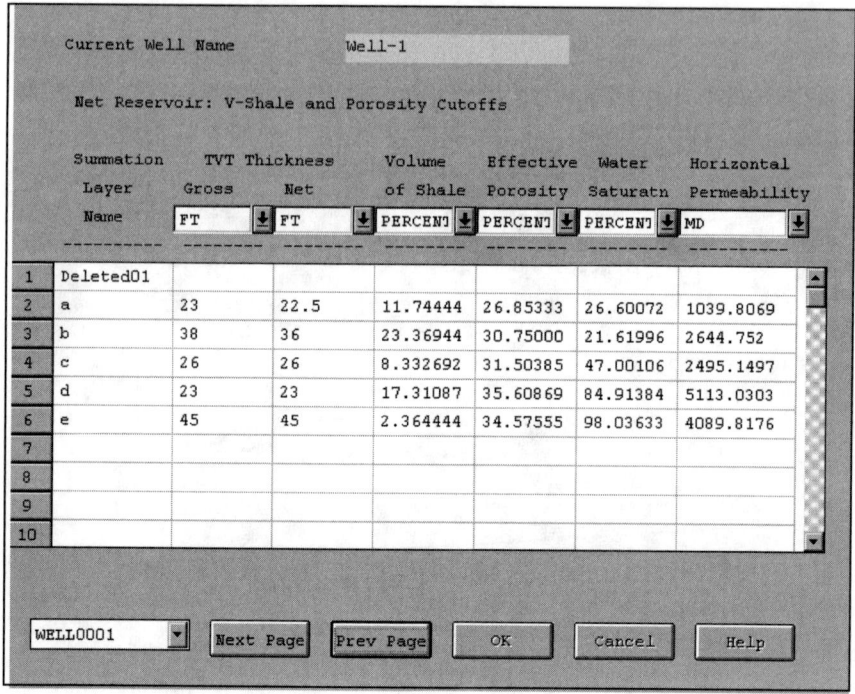

Fig. 5-6 • Summation Results Table

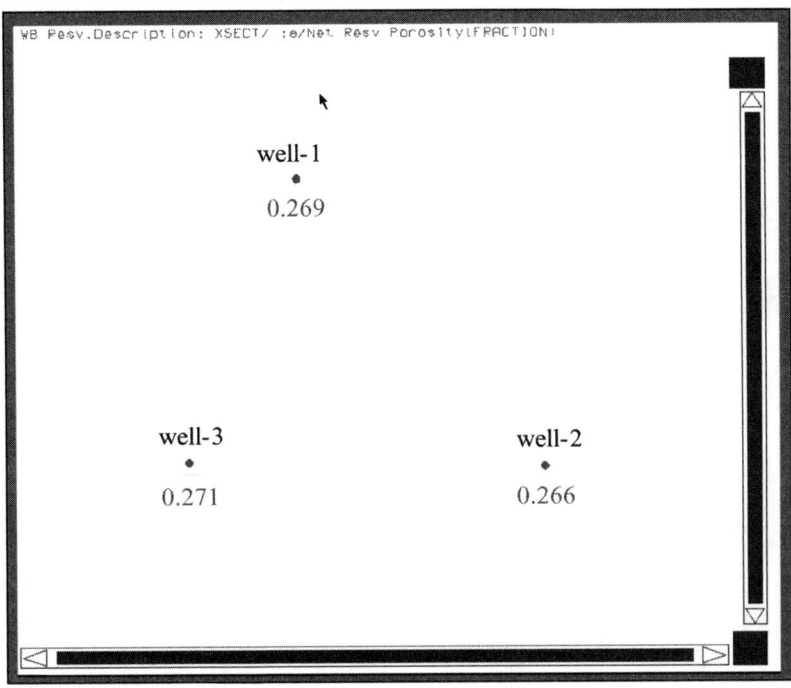

Fig. 5-7 • Posted Porosity Values from Logs

Chapter *SIX*

SEISMIC DATA ANALYSIS

INTRODUCTION

Seismic data, especially 3-D, has gained tremendous importance in reservoir management in recent years. Traditionally, 3-D seismic provided valuable details of reservoir structure and faulting. Improvements in both recording and processing techniques, coupled with advancements in interpretation and visualization software, have revealed seismic expression of heterogeneities in complex reservoirs and even of the fluids within the rocks in some cases.

Information we can usually determine from modern 3-D seismic data includes:

- depth to reservoir
- structural shape, faulting, and salt boundaries
- visualization of reservoir

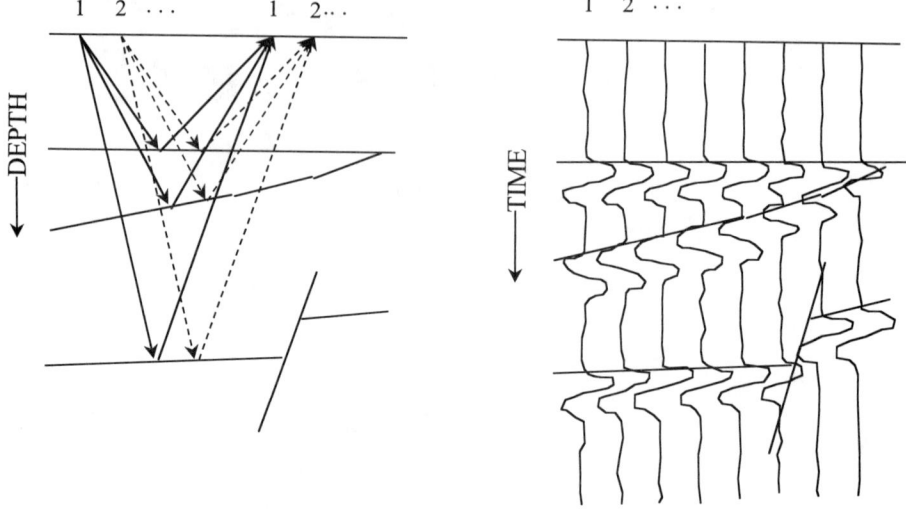

Fig. 6-1 • The Seismic Record Section

In some cases it may also be possible to gain additional insight into:

- porous interval identification
- hydrocarbon reservoir identification
- geologic history
- overpressured zone identification
- properties between wells
- movement of fluids
- fracture orientation

The following discussion briefly explains basic seismic concepts to provide a better understanding of the interpretation.

SEISMIC MEASUREMENTS AND PROCESSING

Seismic reflection occurs at an interface where rock properties change, as illustrated in the depth diagram of Figure 6-1. This simplified example of seismic recording shows the ray path of energy from shot 1 being partially reflected at the top of each rock layer and recorded at receiver 1. The display of the resulting "seismic trace" number 1 is shown on the cross-section of the time diagram in the figure, where each reflection appears as a "wavelet" at the point corresponding to its arrival time at the receiver. The procedure is repeated, with shot and receiver 2 moved a few meters down the line, and so on. (For efficiency in actual field recording, several hundred receivers spread along the line or over an area are recorded simultaneously from each shot. Later, traces with a "common reflection point" are gathered together and averaged to enhance the signal in a single display trace).

The size of the recorded wavelet depends on the reflection strength, $_o$, which is not determined just by the properties within one rock layer, but rather by the *contrast* in acoustic impedance (which equals velocity * density) of the two layers above and below the interface (Fig. 6-2):

Normal Incidence Reflection

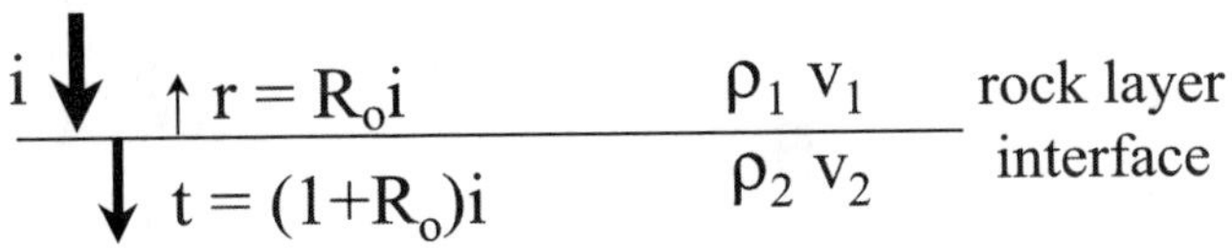

$$
\begin{array}{ccc}
i \downarrow \quad \uparrow r = R_o i & \rho_1\, v_1 & \text{rock layer} \\
\hline
\quad \downarrow t = (1+R_o)i & \rho_2\, v_2 & \text{interface}
\end{array}
$$

$$
R_o = \frac{\rho_2\, v_2 \;-\; \rho_1\, v_1}{\rho_2\, v_2 \;+\; \rho_1\, v_1}
$$

Fig. 6-2 • Seismic Reflection Amplitude

where:

ρ_1 = bulk density of the upper layer

ρ_2 = bulk density of the lower layer

v_1 = acoustic velocity in the upper layer

v_2 = acoustic velocity in the lower layer

R_o = the reflection strength at 0 angle of incidence, *i.e.,* perpendicular to the interface

In Figure 6-2, "i" represents the incident amplitude at the interface, "r" is the reflected amplitude, and "t" is the transmitted amplitude. Their relative values are indicated qualitatively by the sizes of the corresponding arrows.

Figure 6-3 is an example record section in which color represents amplitude. Changes in color along any particular "event" indicate variations in rock or fluid properties, either above or below the interface. If we can assume the layer above the interface is a uniform cap rock, then the amplitude changes can be attributed to the reservoir itself.

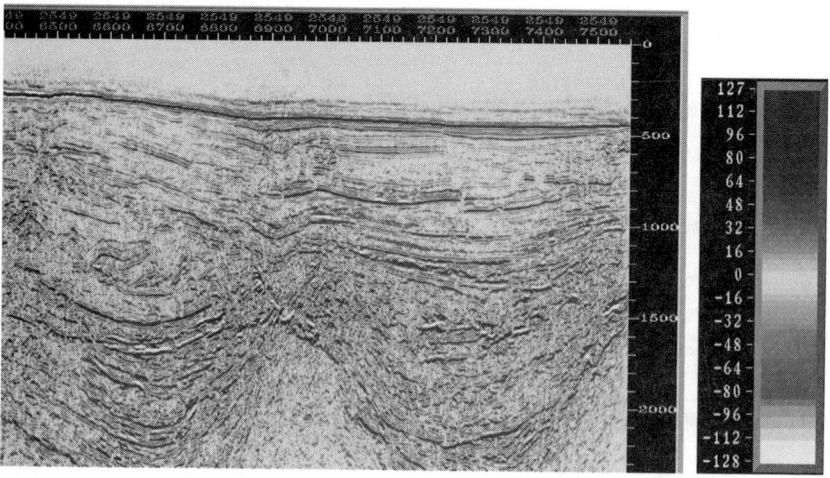

Fig. 6-3 • Seismic Amplitude

It is important to be aware of the vertical resolution in seismic data. The "shot" whose energy is recorded creates a wavelet of finite length. As the wavelet travels through the rocks, high frequency components are attenuated more than low frequencies, causing the wavelet to get longer. By the time it reaches the receivers, it looks something like Figure 6-4. So wavelets arriving from the tops of adjacent thin layers overlap and interfere with each other on the recorded trace, making it impossible to resolve thin layers.

The geometry of seismic ray paths must also be considered. If the rock layers are dipping, the true location of the reflection is not known. The trace is plotted as if all reflections occurred directly below the source-receiver midpoint, as shown in Figure 6-5. A further processing step, called "migration" is required to move the dipping events to their proper locations.

The recorded reflection event is also influenced by its travel through the shallower rocks, which may transmit the signal with little distortion or may alter it severely. So the degree to which reservoir properties are visible

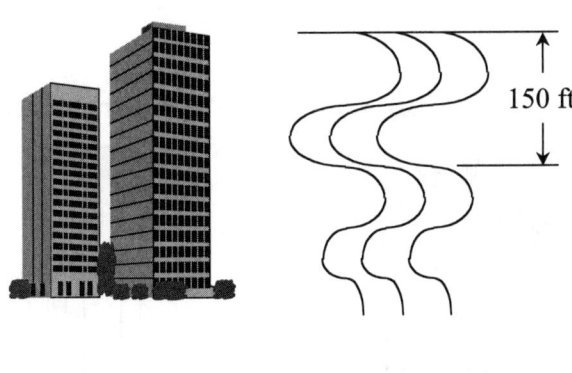

$$\lambda = \frac{v}{f} = \frac{10000\,ft\,/\,\sec}{67\,cyc\,/\,\sec} = 150\,ft\,/\,cyc$$

Fig. 6-4 • Vertical Resolution

in seismic data varies with the geologic setting. When conditions are right, a significant contribution to the reservoir description can be gained from seismic data.

STRUCTURAL INTERPRETATION

Figure 6-6 shows a typical seismic cross section from an offshore area. The horizontal axis is map position. The vertical axis is 2-way travel time for the seismic waves (sort of a non-linear depth scale). The color or shade of gray indicates the strength of the reflected signal. A lot of information about the geology is apparent to the trained eye.

Figure 6-7 shows the same data with some faults interpreted and the outline of a salt dome indicated. By correlation with well logs (if available), the interpreter will determine which of the seismic events corresponds to the

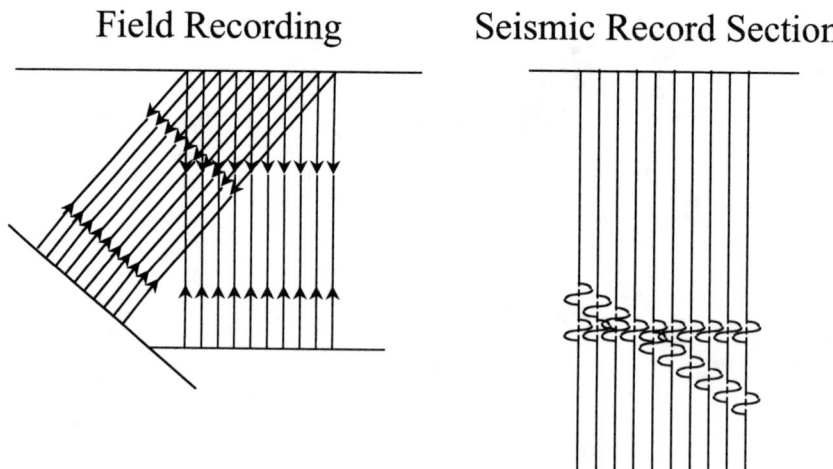

Field Recording   Seismic Record Section

Fig. 6-5 • Migration

reservoir and follow it throughout the 3-D volume, producing a structure map of the top-of-reservoir. This map is usually plotted in seismic travel time, but may be converted to depth using velocities extracted from special velocity surveys and/or from the raw seismic data.

Conventional seismic data processing makes the simplifying assumption that the rocks are composed of horizontal layers, and that the reflection point in the subsurface is half way between the seismic source and the receiver. When this is approximately the case, seismic traces can be processed and displayed as a function of seismic travel time. Traces having a common reflection location, but a different source-to-receiver distance, are "stacked" (summed to reduce noise) to image the rock layers clearly, without having to first create a detailed model of the rock velocities as they vary both horizontally and vertically.

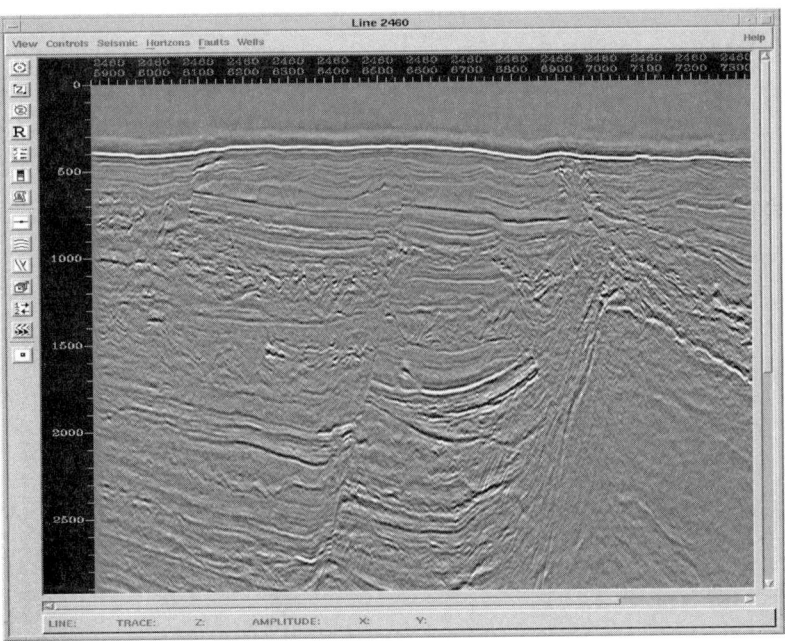

Fig. 6-6 • Seismic Cross Section

When there are gradual structural changes in the rocks, these assumptions still work reasonably well. An additional process called *post-stack time migration* is then required to make adjustments in structural positions and shapes, as discussed in Figure 6-5.

In areas of complex geologic structure, the assumptions used in conventional processing do not hold. This causes the images to be poorly focused during the stacking process. They look "fuzzy" and/or dipping reflection images are not positioned correctly. To solve this problem, geophysicists have developed a method for converting the data from seismic travel time to depth and doing the migration before the stack. This *pre-stack depth migration* requires an iterative procedure of estimating a model of the rock velocities and using it to image the data. Then using the image to give a better

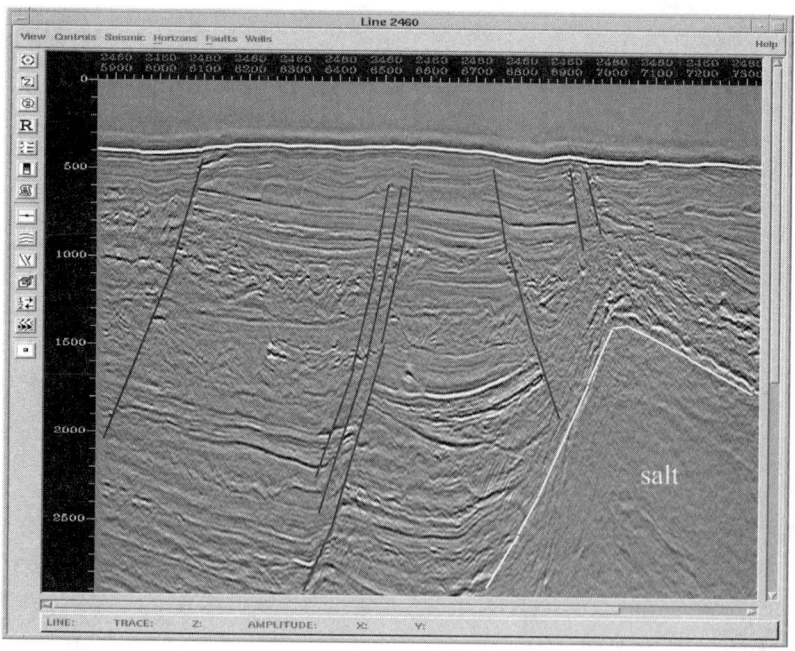

Fig. 6-7 • Seismic Interpretation

estimate of the velocity model, which is used to get a better image, and so on. When the model is correct, the pre-stack depth migration image is sharply focused and reflections are properly positioned.

Figure 6-8 shows a comparison of time migration and depth migration. Note the vertical scale difference and the non-linear stretch of the time section due to velocity variations. Such improvements in seismic images make it easy to understand why pre-stack depth migration has become the norm in complex areas.

Post-Stack Time Migration

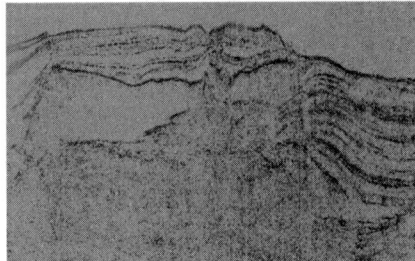

Pre-Stack Depth Migration

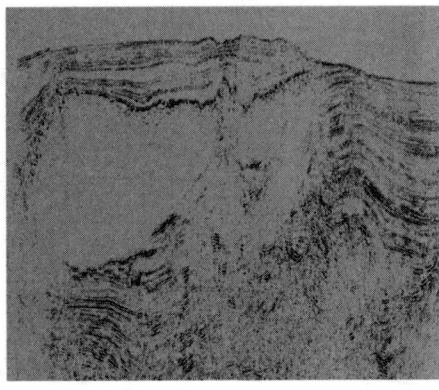

Fig. 6-8 • Time Migration vs. Depth Migration

STRATIGRAPHIC INTERPRETATION

Once a seismic horizon has been interpreted, its reflection strengths (also called seismic amplitudes) can be displayed as a map. The example in Figure 6-9 is from a shallow horizon. It shows evidence of a buried stream channel, which was filled with sand or mud having different acoustic impedance than the surrounding rocks. Whenever a heterogeneous rock layer is

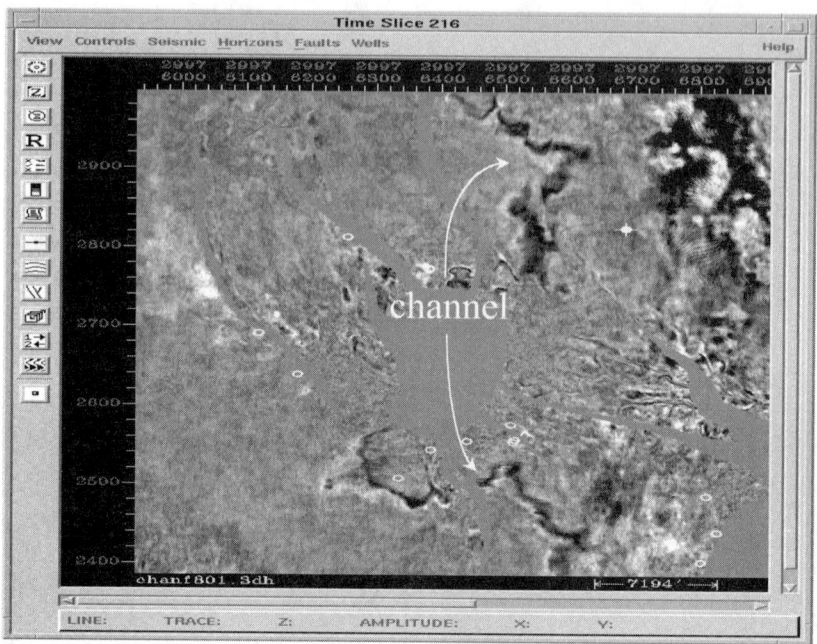

Fig. 6-9 • Seismic Amplitude Map

overlain by a uniform layer, there is the possibility that the changes in density or acoustic velocity will be visible in such a seismic reflection strength map.

Variations of porosity and of the fluids filling that porosity also alter the acoustic impedance of the reservoir rock. So anomalies in rock porosity or fluid content can sometimes be detected indirectly in seismic amplitude maps. Such maps can be used directly to position future wells. They can also be used to guide the characterization in a reservoir model.

3-D VISUALIZATION

Three-dimensional visualization is a tool for displaying 3-dimensional data on a 2-dimensional screen. The data can be rotated, panned, and zoomed. Motion often makes the shapes easier to understand. Seismic surveys, geostatistical 3-D grids, and simulation models can be loaded into the software to enhance seismic interpretation and reservoir management.

Commercial 3-D visualization software allows the user to interactively examine an entire volume of data in one image. Starting with an opaque view of a complete volume, the user can adjust the transparency to visualize the spatial distribution of selected attributes (*e.g.*, amplitudes) and extract structural relationships from the data set. Alternatively, an interpreter can choose a range of values for an attribute, select one "seed point," and have the software "auto-pick" all connected data with similar values. An example of auto-picking is shown in Figure 6-10, where part of the seismic data has been stripped away after two horizons were picked. Auto-picking and 3-D

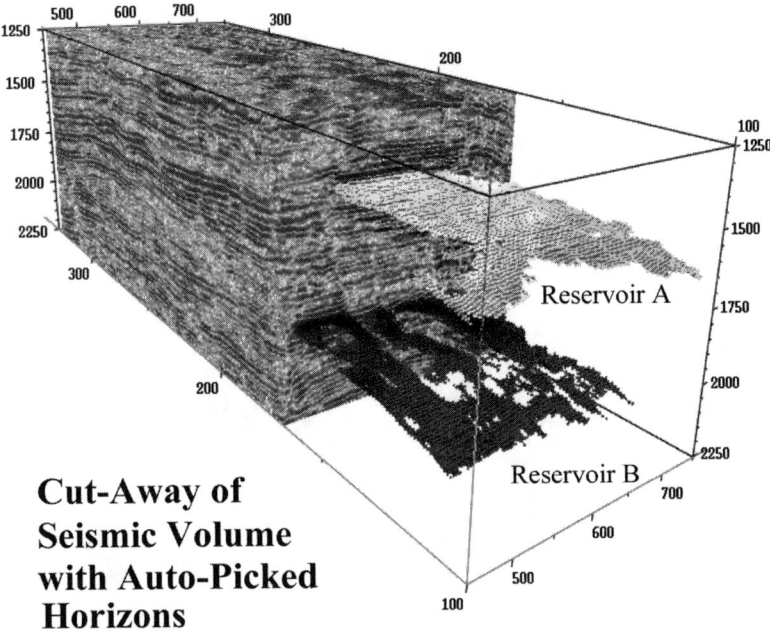

Cut-Away of Seismic Volume with Auto-Picked Horizons

Fig. 6-10 • 3-D Visualization

visualization cannot only decrease interpretation time by orders of magnitude, but can also greatly improve the interpreter's understanding of the reservoir, revealing detailed information about reservoir quality and heterogeneity.

REFERENCES

Suggestions for further reading:

- Brown, Alistair R.: *Interpretation of 3-Dimensional Seismic Data,*
AAPG Memoir 42
- Jenkins, Waite, and Bee, *The Leading Edge,* Sept, 1997

Chapter SEVEN

MAPPING AND DATA VISUALIZATION

INTRODUCTION

For many years, computer software has provided tools for combining and displaying data from geologic studies, seismic surveys, and well logs. Maps of spatial data include important auxiliary information, such as lease boundaries and cultural descriptions. Various plots, charts, and maps help the interpreter better understand the results of decline curve analysis and reservoir simulation, such as well productivity and fluid saturations.

Computerized mapping programs are commonly available and can save a great deal of time, as well as enhance accuracy of display and understanding of the data. A word of caution is in order, however. While fully automatic maps using default parameters can be handy for a quick look and data editing, they are generally not suitable for model description in simulation or for reserve calculations.

Contouring algorithms are only as good as the data supplied to them. They cannot incorporate critical conceptual information, which is often needed to make the results "geologically reasonable."

Every computer-generated map should be carefully checked. Manual editing should be done as needed to ensure that the map is correct and appropriate for its intended purpose.

CONTROL DATA

The manual editing needed to prepare acceptable maps is done by adding "control data" to guide the computer contouring (especially at map edges). Several types of control data can be entered:

- Well values—values of a property in a layer at a well spot
- Control contours—digitized from a paper map, or drawn freehand on the screen by the user to change the generated contour
- Control points—like control contours, but single points
- Faults—digitized or drawn

DATA ENTRY

Data can be input to mapping programs in several ways. These include:

- digitizing
- importing files
- keying in
- screen editing, using a mouse

MAPPING SOFTWARE EXAMPLE

Some of the features of a typical mapping program are illustrated in the following figures, which were created with Baker Hughes/SSI's Petroleum WorkBench.

Display items control

The display can usually be thought of as an overlay of many layers of information. Options allow viewing of selected items, properties, and model layers, such as:

- Surface components
  - well surface spots
  - well names
  - base map lines
  - base map text

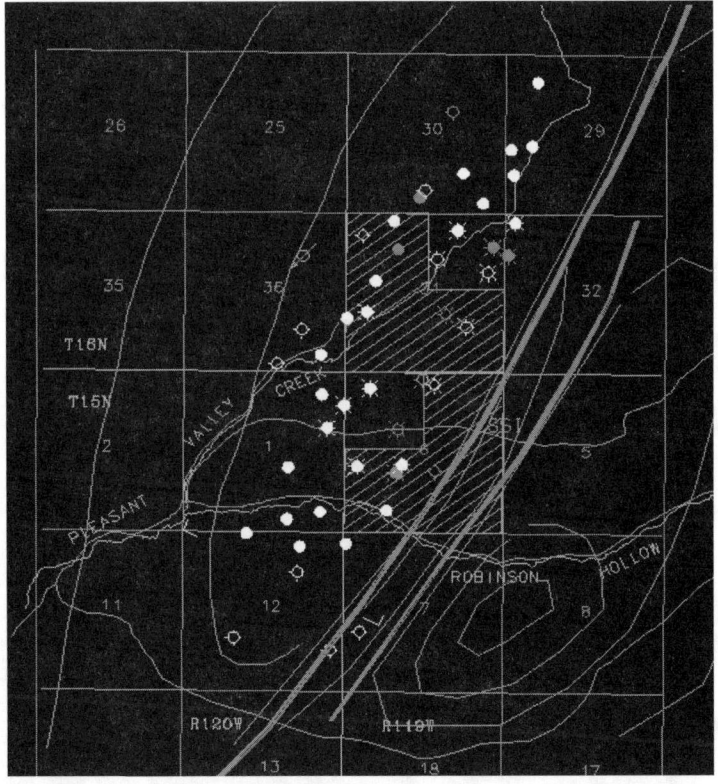

Fig. 7-1 • Base Map with Structure Control

    - base map scale
    - lease lines/names
    - grid (volumetrics, simulation)
- Layer components
    - well layer spots
    - well values
    - control contours
    - control points
    - faults
    - generated contours
- Grid cell values (show numbers or paint color)

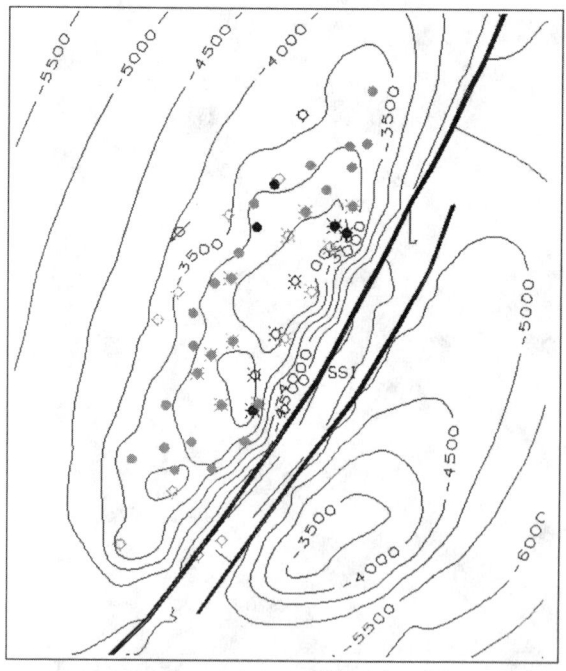

Fig. 7-2 • Computed Structure Contours

Full zoom control is also generally available. Figure 7-1 illustrates some of the types of data commonly included in a computer-generated map.

MODEL DATA

Control data consisting of array data, input points, input contours, faults, and/or well values are used as the basis for contouring. Unless the user specifies otherwise, faults are used only when contouring "structure-dependent" properties. Figure 7-2 shows structural contours computed independently in each fault block from the input control data.

ALTERNATIVE DISPLAYS

Once the map values have been calculated, there are several ways to display the results. The contours shown in Figure 7-2 are similar to a hand-made map. But with computer software, there are further capabilities. Various color-fill displays of such 2-D maps are common. Alternatively, a 3-D display, such as Figure 7-3, may help to visualize the surface. Such displays

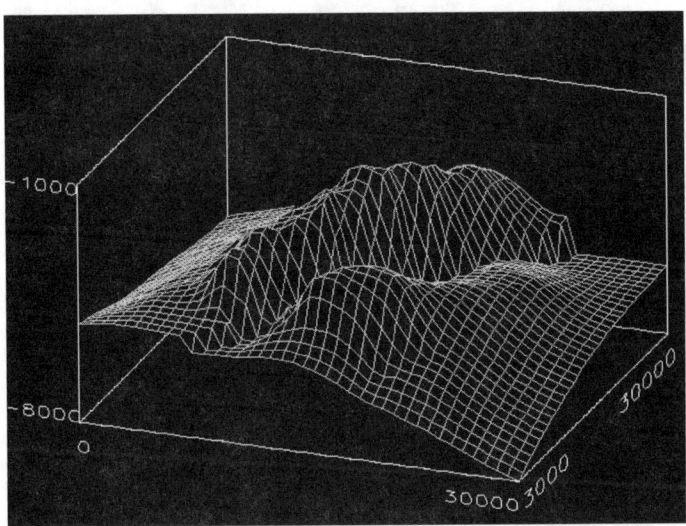

Fig. 7-3 • 3D Mesh Plot

can be tilted and rotated. A so-called 4-D display is shown in Figure 7-4. Here the mesh shows the *structure* in 3 dimensions, while the addition of color allows simultaneous display of *porosity*.

Many types of computer display are available. Recently, dramatic increases in workstation computing speed and data capacity have allowed the development of "3-D visualization" software that was the stuff of dreams only a few years ago. Geoscientists and engineers, working with software developers, are inventing new techniques for "seeing" what the data have to reveal about the nature of hydrocarbon reservoirs and their behavior under various operating scenarios. With these new capabilities, speed of interpretation is increasing by orders of magnitude, with concurrent increases in depth of understanding.

One dramatic example of these developments is in 3-D seismic data interpretation. In some reservoirs, increases in the amplitude of seismic

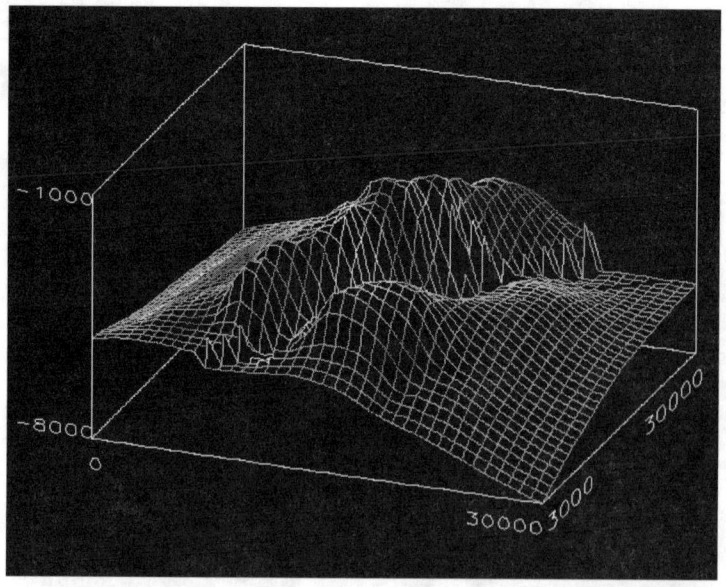

Fig. 7-4 • 4D Mesh Plot

reflections are directly related to the presence of hydrocarbons. The new software allows instantaneous selective viewing of the data. By displaying only the highest amplitude data and rendering all other data "transparent," the interpreter can immediately see the shape and extent of the hydrocarbon accumulation within such a reservoir. The same technique can be used with the output of a reservoir simulator to visualize the distribution of fluids throughout the reservoir at various stages of production.

Chapter *EIGHT*

GEOSTATISTICAL ANALYSIS

INTRODUCTION

The mapping described in chapter 7 is a simple extension of the maps one would make by hand contouring. An alternative procedure measures statistical variations of the data points and then creates maps having similar statistical properties throughout. This procedure is called "geostatistics," because it was first used in the mining industry to map geological properties.

Like conventional mapping, geostatistics seeks to answer the question "What are the reservoir property values between the wells?" (Fig. 8-1). But analysis of the statistical distribution of data values can lead to more detailed estimates of map values between measured points, better describing the true heterogeneity of the reservoir.

The usual measure of this statistical variation is called a "variogram." It is a mathematical expression involving the measured data points, which represents how the data values change from one location on the map to another. The amount of change is measured as a function of the distance between data points, and sometimes as a function of direction. The variogram serves as a tool for interpolating the map property between known points, while preserving the degree of variation seen in the data.

Another benefit of geostatistical mapping is that it provides a means for displaying the *uncertainty* of the interpolated values—an important piece of information that is not available from conventional maps.

A third advantage of geostatistical mapping is that it can integrate independent measurements of a map property. For example, if both well logs and 3-D seismic amplitude are related to reservoir porosity, a map of porosity can be made that ties the wells together and also incorporates the trends between wells that are seen in the seismic data. Thus the correlation of the "hard data" of the porosity log and the "soft data" of the 3-D seismic are integrated to create the porosity map.

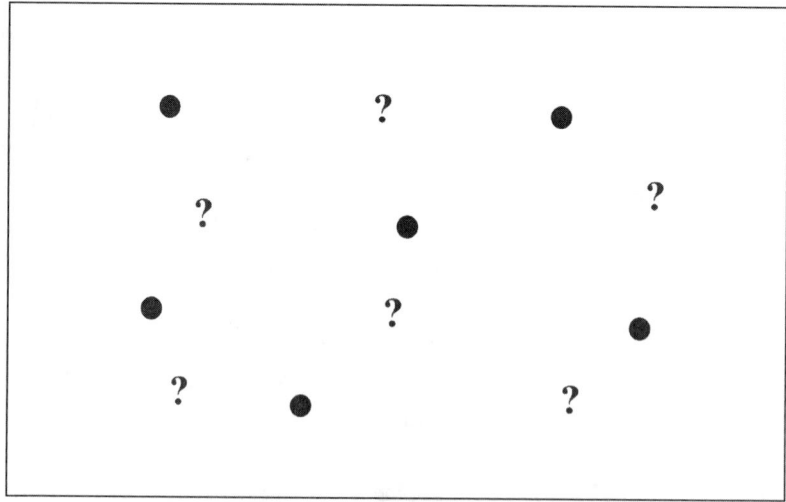

Fig. 8-1 • Geostatistics- Reservoir Properties Between the Wells?

The following discussion provides an overview of the geostatistical method, illustrating the concepts mentioned above by showing a real oil reservoir example:

CONVENTIONAL MAPPING

Well log analysis generally provides the measured data that serves as the basis for mapping reservoir properties such as:

- gross thickness
- net thickness
- porosity

In the Rocky Mountain example of Figure 8-2, two formations are identified in a cross-section through a few of the wells.

Analysis of all available logs throughout the field resulted in the map of net reservoir thickness shown in Figure 8-3. The task at hand is to estimate the net thickness at all locations between and surrounding the wells.

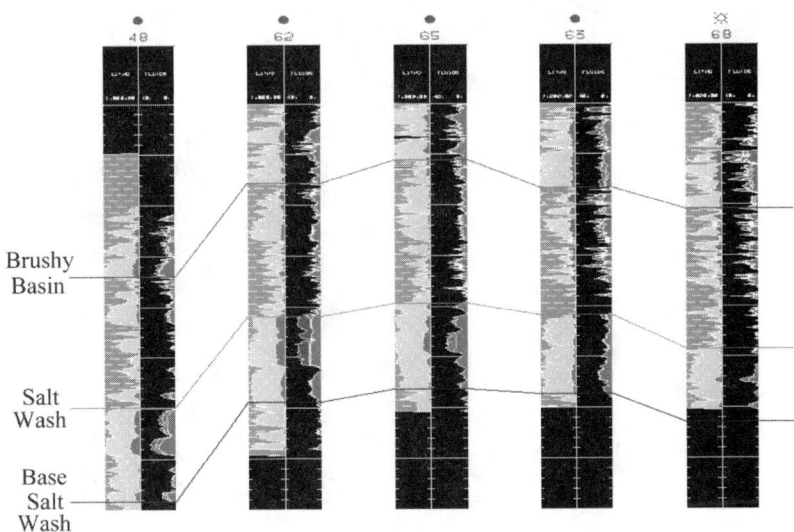

Fig. 8-2 • Well Logs Providing Reservoir Properties

The conventional technique for estimating values is to contour the data, by hand, or by using a computer algorithm. If hand-contoured by a geologist, the resulting map can incorporate knowledge of the geologic trends and depositional patterns. But maps made by 10 different people would likely give 10 somewhat different thicknesses at any point that is not fairly near a well.

Mathematical algorithms can provide consistent means of contouring the data, with results like those of Figure 8-4. Such algorithms generally cannot incorporate geologic character unless provided with additional "control" points and/or contours by the geologist.

MAP STATISTICS

Geostatistics provides a means of interrogating the data to determine how a reservoir property varies with distance and direction from any given

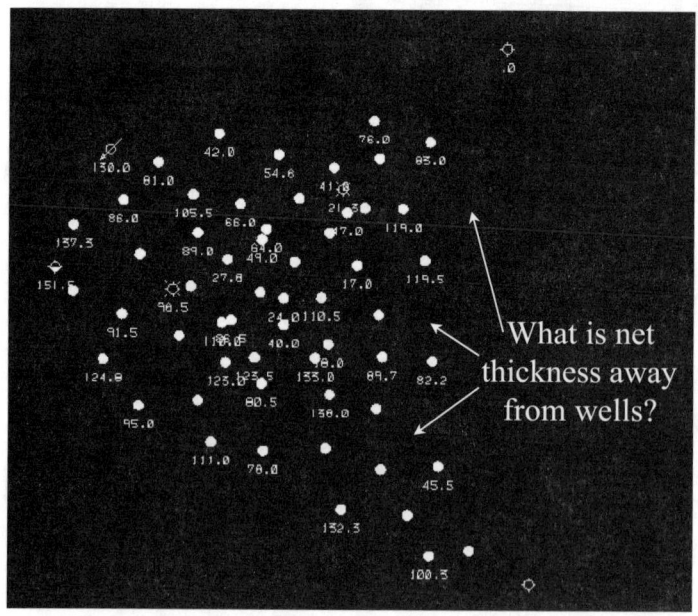

Fig. 8-3 • Salt Wash Net Thickness Measured at Wells

point. A map that includes these trends can then be created. Net thickness is used in Figure 8-5 and the following examples to illustrate the concepts, which apply similarly to any property that can be mapped.

The common measure of variation with distance and, optionally, direction is shown in Figure 8-6. It is called a *variogram*, or to be precise, a semivariogram, due to the 1/2 in the equation:

Here $Z(x)$ is the net thickness at one well and $Z(x+h)$ is the net thickness at a well a distance h away. The difference in net thickness is squared and added to the squared differences in all other pairs of wells that are separated by distance h. Since wells are not usually spaced on uniform increments, distance of separation is divided into "bins," such as the donuts shown in the figure. Then "h" represents any distance within the donut. The summation is over the "n" pairs of wells found at this distance. Dividing the sum by $2n$ gives the average variance.

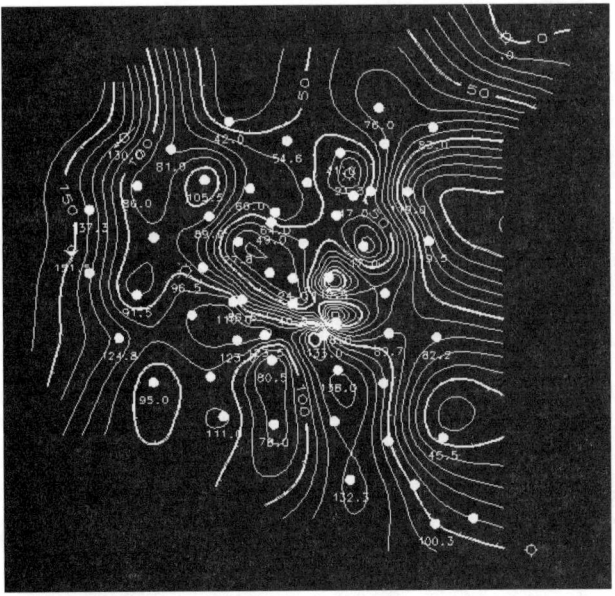

Fig. 8-4 • Conventional Contouring

GEOSTATISTICS uses the spatial correlation of measured values of a property to estimate the value of the property of other locations.

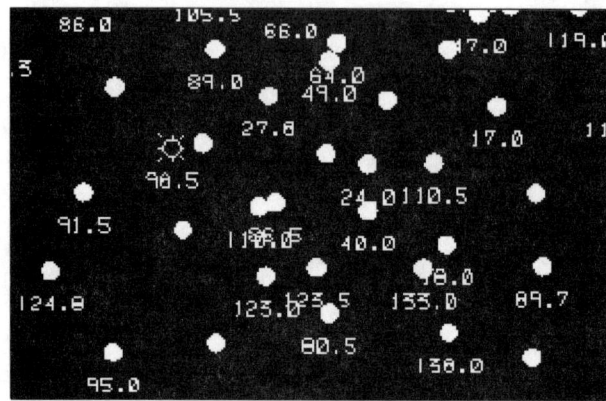

How does the net thickness vary with <u>distance</u> and <u>direction</u> from one well to another?

Fig. 8-5 • Spatial Correlation of Data

The result calculated for net thickness for the *n* pairs of wells in the donut represented by distance *h* is plotted in Figure 8-7. The process is repeated at all other distances to complete the variogram.

Several important terms describing characteristics of the variogram are defined in Figure 8-8.

When wells are very close together, we expect their net thicknesses to be similar. So, as *h* approaches zero, the variance should approach zero. The variance at *h*=0 is called the *nugget*. For the types of data normally used in reservoir studies, the nugget should be zero. If it is not, then there are not enough measurements at short distances or there are measurement errors.

As well separation increases, the variance of net thickness typically increases until a distance is reached where there is no longer any statistical significance. At this distance, called the *range*, the variance tends to flatten.

The variance at the range is called the *sill*.

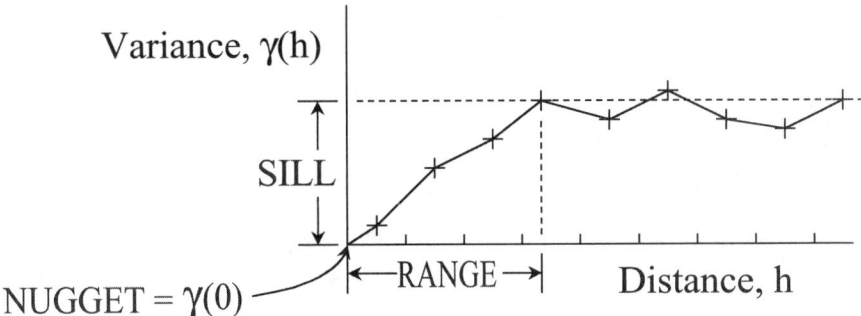

Fig. 8-8 • Variogram Terminology

from the wells, where the conventional contouring does not indicate the geological trends measured by the variograms and incorporated in the kriged map.

An alternative to contouring is to represent the map values by colors (Fig. 8-13). This also reveals the size of the grid used for the kriging calculations.

Regardless of the method used to create the map, uncertainty in the estimates increases with distance from the wells. The four measured points shown on the vertical cross-section of Figure 8-14 represent well values, and the double-headed arrows represent the range of possible values at locations between the wells.

The line through the middle is the kriged value. Kriging creates a smooth map that ties the wells and estimates the most likely values between the wells.

What kriging does not show is the range of possible values between the wells, shown in the figure by the two bounding lines.

SIMULATION

An alternative to the single map produced by kriging is to generate many maps spanning the range of possible estimates between the bounding

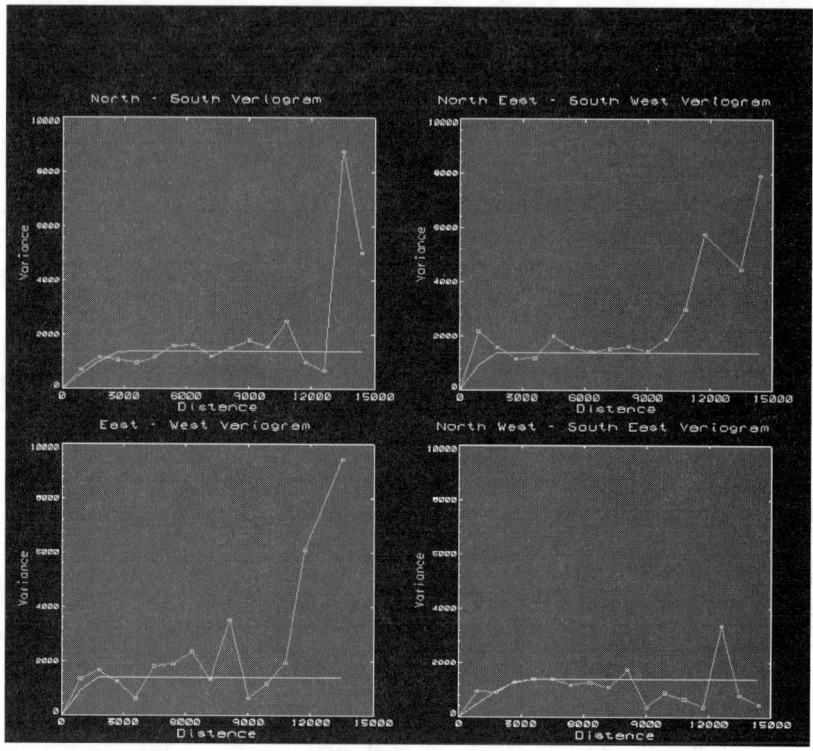

Fig. 8-9 • Directional Variograms

lines. This is called *simulation*. Simulation creates a mathematical model of the mapped property that has the same spatial statistics as the actual data. There are many possible maps that differ from each other, but have the same statistics. Some of them meet the additional condition that they honor the measured data points (*i.e.*, the maps tie the wells). In the usual case where the simulation is forced to honor known data points, the process is called *conditional simulation*.

Each individual simulation map, called a *realization*, is one of many possible estimates. Maps produced by simulation include more fluctuations

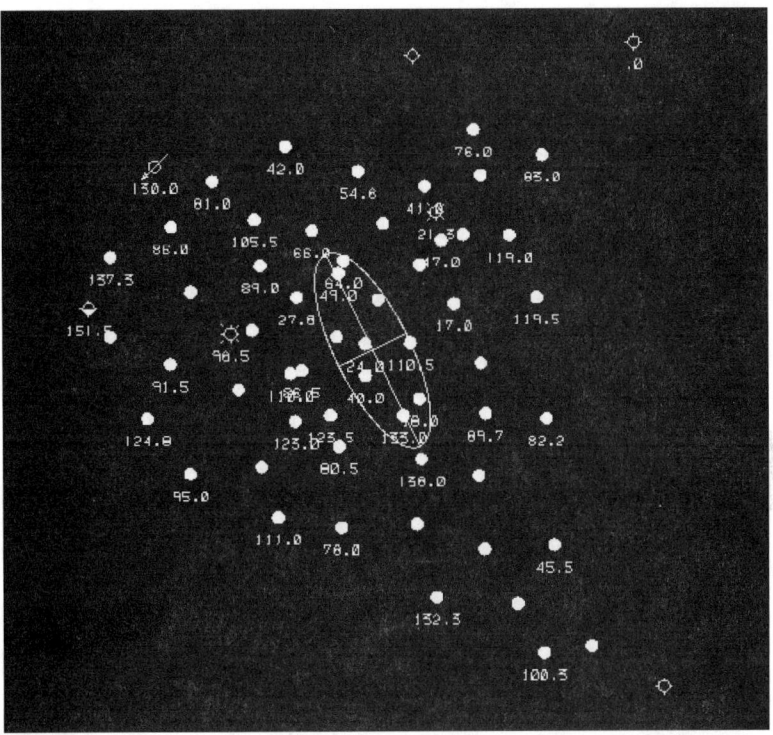

Fig. 8-10 • Variogram Ellipse

than kriged maps, because they allow for the extremes in the estimated values. No single map represents the "most likely" case. Instead, the set of all realizations represents the range of possibilities. Each of the realizations is equally likely to represent the actual reservoir condition.

In comparing maps produced by these different methods, note that:

• kriging estimates represent a moving average process, and the resulting map is heavily smoothed
• simulation produces a series of more rapidly fluctuating maps,

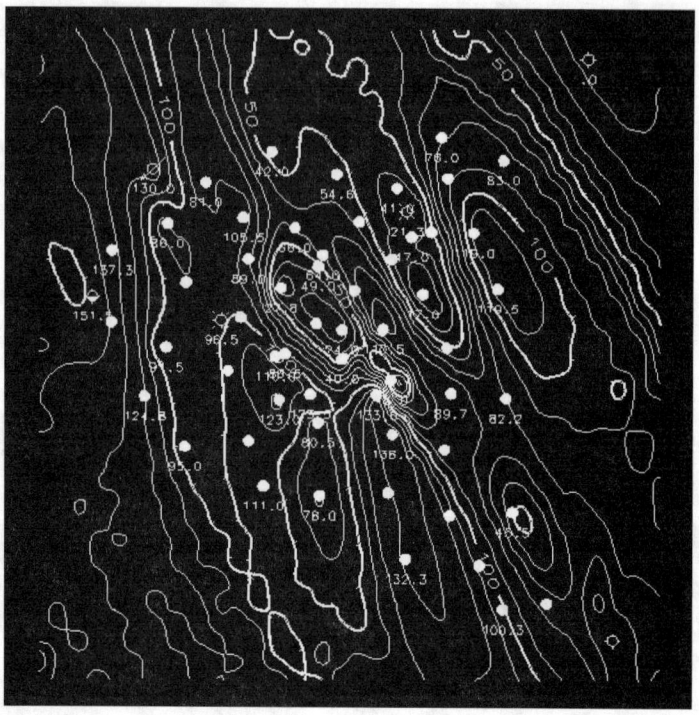

Fig. 8-11 • Kriged Map (Contoured)

which all have the same general features, but exhibit the range of possibilities

These concepts are most easily understood by looking at an example. In Figure 8-15, the smooth kriged map is compared to some of the nine realizations from a simulation. The number of realizations is arbitrary. The color scale is the same for all maps, making it clear that the simulations include more extreme possibilities among the predicted values between the wells. Any of the realizations is probably a better representation of the texture of the actual geology, while the kriged map represents the "most likely" value at any particular location.

Kriging Reflects Data Trends

Conventional Contouring   Kriged Contouring

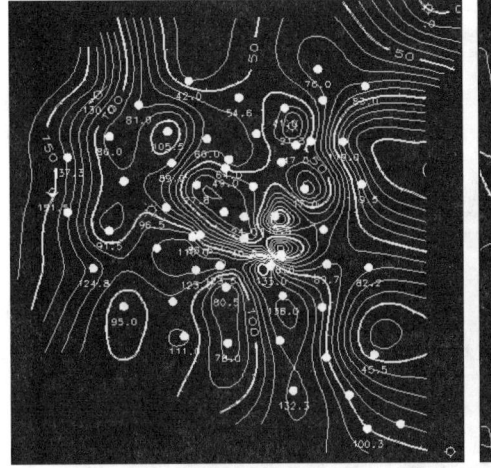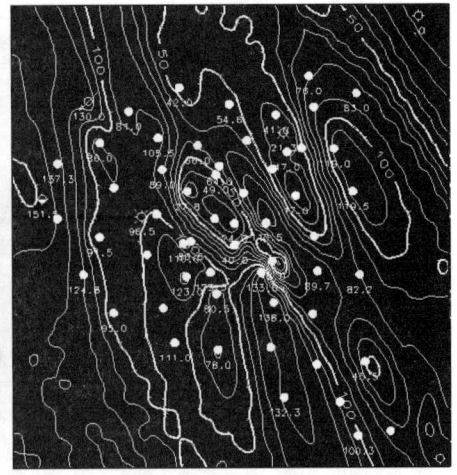

Fig. 8-12 • Kriging Reflects Data Trends

MEASURING UNCERTAINTY

Geostatistics provides an opportunity to measure the uncertainty of the maps it produces. The kriged map of Figure 8-16 predicts a relatively large net thickness in the area of a shallow well that did not penetrate this reservoir. It has been proposed to deepen the well and complete it here. But how certain are we of the net thickness shown on this map? The following example shows how simulation can be used to determine the uncertainty in the predicted net thickness and, thus, the risk associated with drilling this well.

The plot on the right in Figure 8-17 is the *probability density function* (PDF) from the simulation for the grid cell containing the proposed well. It shows that of the nine realizations, two were estimated with a net thickness of 107 to 117 ft, three were estimated 118 to 128 ft, and four were estimat-

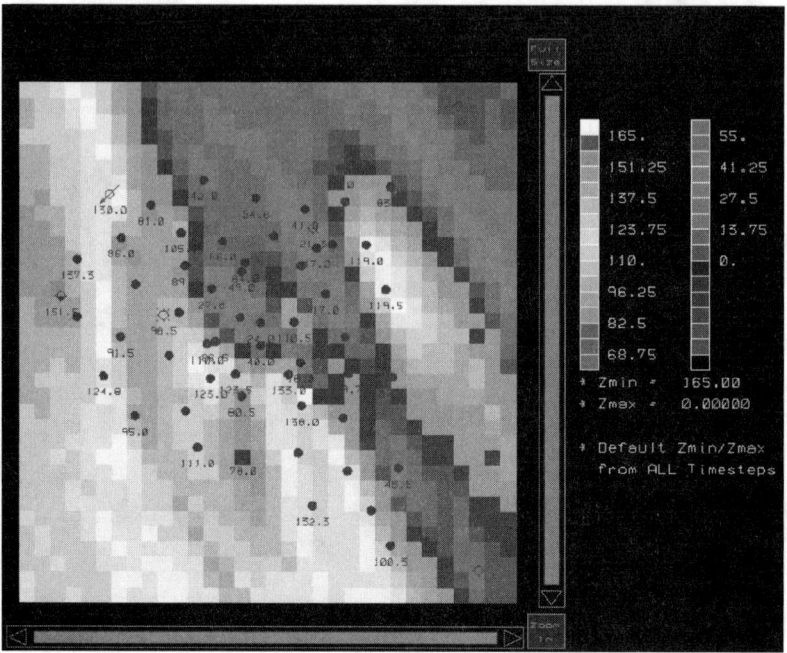

Fig. 8-13 • Kriged Map (Color Filled)

ed 129 to 140 ft. (Using a larger number of realizations would have resulted in smaller bin sizes here, and a smoother histogram).

The plot on the left is the cumulative probability distribution, also called *cumulative density function* (CDF). It is simply a cumulative graph of the PDF. The vertical axis gives the "probability" that the net reservoir thickness will be less than the corresponding value on the horizontal axis. For example, there is a 35% probability that the net thickness will be less than 120 ft and a 90% probability that it will be less than 132 ft.

These plots represent only the grid cell at the proposed well location. An alternative display, shown in Figure 8-18, gives the probability of finding net thickness greater than 125 feet anywhere on the map. White areas have no chance of thickness this large. The proposed location, shown by the dark

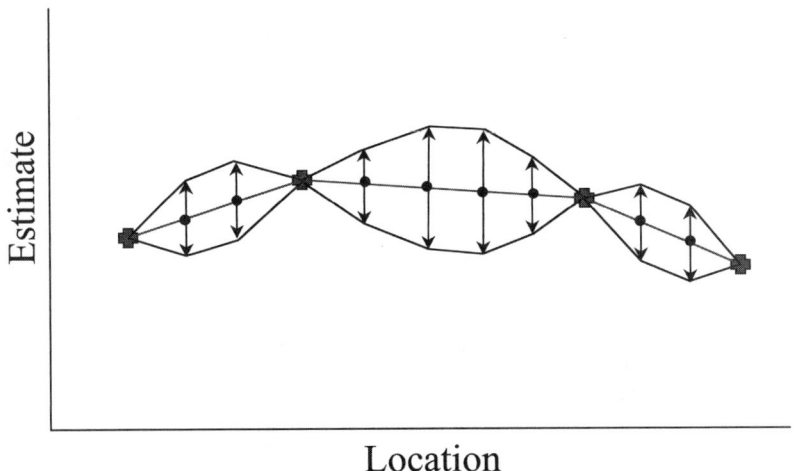

Fig. 8-14 • Uncertainty in Kriging

well symbol, is again seen to have a good chance of success, with about a 55% chance of finding more than 125 feet of pay.

EXTENSION TO 3 DIMENSIONS AND TO CO-KRIGING

Geostatistics can also:

- estimate variation in 3 dimensions
- integrate additional measurements as either hard or soft data

The examples discussed above use geostatistics to create maps in 2 dimensions. The same techniques can be extended to 3 dimensions. For

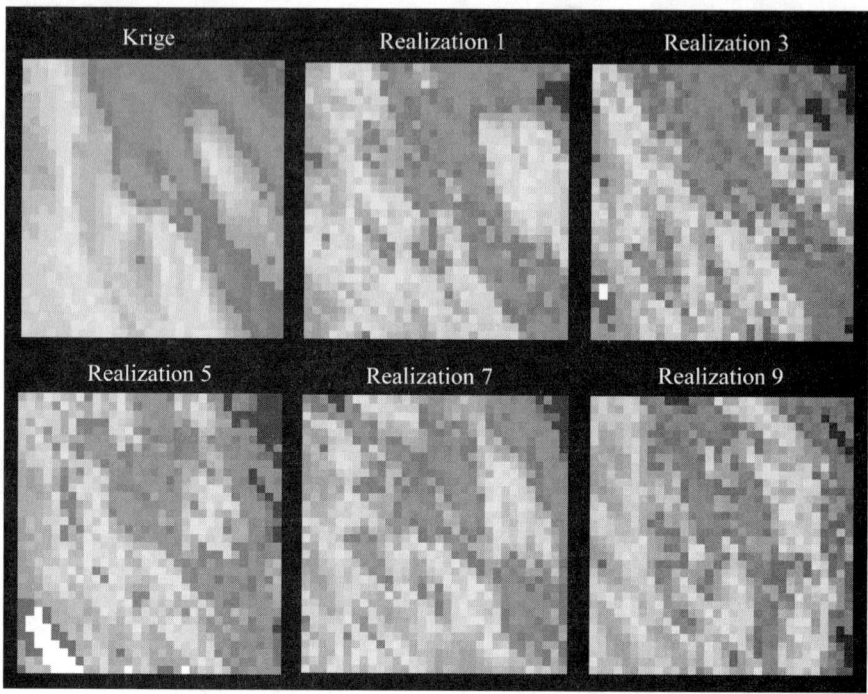

Fig. 8-15 • Realization Comparisons

example, well logs provide measurements of reservoir properties over an interval of depth. In mapping porosity, a vertical variogram can be calculated, to determine the variance of porosity with sample separation in the logs. The techniques described above can then make use of both the vertical and horizontal variograms to populate a 3-dimensional grid with porosity estimates by either kriging or conditional simulation. Figure 8-19 is an example of a 3-dimensional geostatistical model created from temperature logs during a steam flood.

Sometimes additional indicators of a reservoir property may be available. Seismic amplitude might correlate to reservoir porosity, for instance. Since seismic is an "indicator" of porosity, but not an absolute measure, it can be

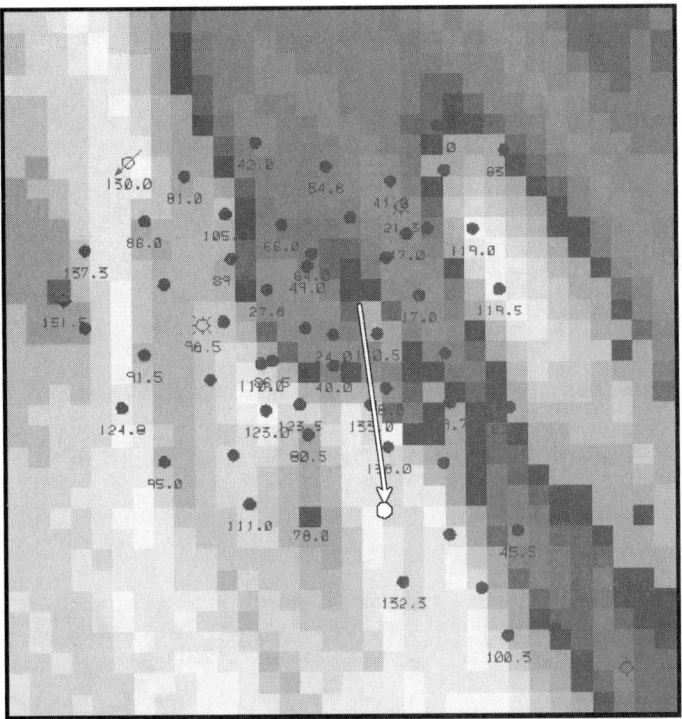

Fig. 8-16 • Predicted Net Thickness

treated as "soft data" in geostatistical calculations. When the "hard data" at the wells are used to generate a porosity map, the estimates away from the wells are modified somewhat to reflect the suggested trends from seismic amplitude. This can be a very powerful tool for reducing the map uncertainty, and it is especially useful to estimate values between widely-spaced wells.

CONCLUSIONS

Geostatistics provides:

• a means of estimating reservoir properties that incorporates measured trends

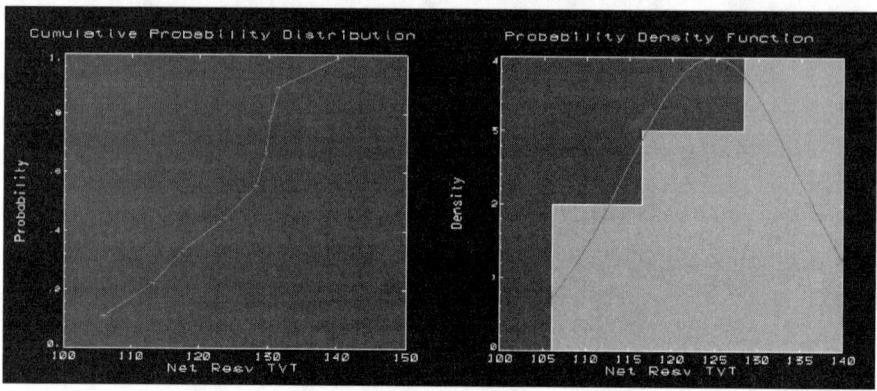

Fig. 8-17 • Probablility Plots for Proposed Well Location

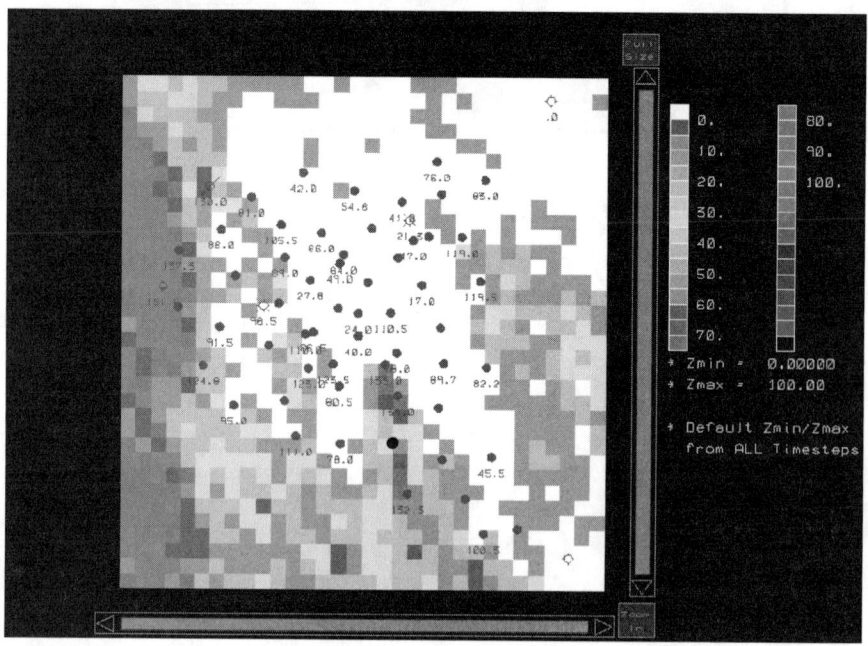

Fig. 8-18 • Probablility of Net Thickness, > 125 feet

Reservoir Temperature After Steam Injection

Fig. 8-19 • 3-Dimensional Temperature Model

- a measure of the uncertainty in predictions
- a method for integrating independent measurements of a reservoir property

We have seen that geostatistics offers 2- and 3-dimensional methods for determining likely property values away from points of measurement. These values are estimated by using the geological trends found in the measured data.

The statistics of the simulation process provide a means of determining the uncertainties involved in predicting reservoir properties. These uncertainties can be visualized by plotting numerous "realizations" of the map, by making probability maps of a particular scenario, and by plotting the calculated probability distribution functions.

If there are multiple types of measurements for the same property, such as well logs and seismic attributes, they can be integrated in the map-making process.

REFERENCES

1. DeLage, Jack, Scientific Software-Intercomp, Houston, Texas, personal communication

Two good textbooks on this subject are:

- Hohn, Michael E.: *Geostatistics and Petroleum Geology*, Van Nostrand Reinhold (1988)
- Isaaks, Edward H. and Srivastava, R. Mohan: *An Introduction to Applied Geostatistics*, Oxford University Press (1989)

Chapter NINE

WELL TEST ANALYSIS

INTRODUCTION

Well pressure transient testing includes the techniques for generating and measuring time variations of pressures in wells, while flowing or shut-in. The well test data are used to determine:

- Reservoir properties—permeability and porosity
- Pressure—bottom hole flowing, and average reservoir
- Wellbore condition—wellbore storage, skin, completion, and flow efficiency
- Boundaries—reservoir geometry and size, sealing faults, pinch outs, and edge-water contacts
- Interference from other wells

The test results are utilized for exploration, reservoir engineering, and production engineering activities.

This chapter will provide fundamentals of well testing, and some insight into well testing software, with examples.

TEST TYPES

Pressure tests in wells usually consist of:

- drawdown and buildup tests on a production well
- fall-off tests in an injection well
- interference tests involving two or more wells, *e.g.*, an injector and surrounding producing wells
- drillstem tests on temporarily completed new wells

As fluids are produced at a constant rate in a drawdown test (Fig. 9-1), reservoir pressure decreases in a manner that depends on reservoir properties

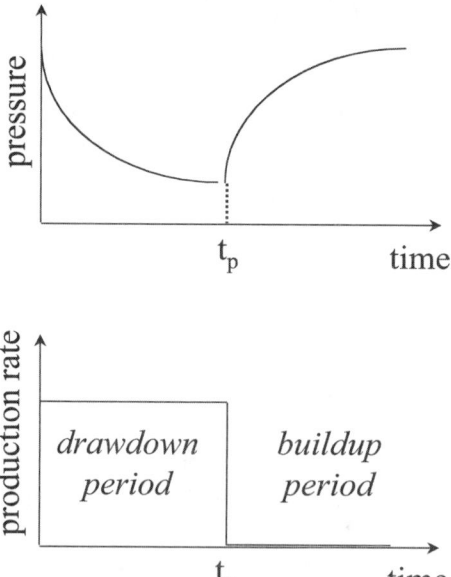

Fig. 9-1 • Drawdown and Buildup

and conditions in and near the wellbore. Advantages of drawdown tests include no loss of production during testing and the potential for determination of oil-in-place with extended tests. The initial pressure, p_i, is an important reservoir parameter that is measured directly. Disadvantages include the requirement that the rate be held constant, which is impractical and rarely achieved.

For a buildup test, the well is shut in and the pressure begins to return to normal (Fig. 9-1). Advantages include precise control of rate (zero) and the ability to determine p^*, the "extrapolated pressure." Disadvantages include lost (delayed) production due to the required shut in.

In an *injection test,* pressures are measured while injecting fluids into the reservoir. In case of a *fall-off test,* the well is shut in after a period of constant injection, measuring the drop of pressure with time. These tests correspond to the producing well buildup and drawdown tests.

A *drill stem test* prior to well completion consists of a series of drawdown and buildup tests to sample the formation fluid and establish the probability of commercial production.

An *interference test* consists of measuring pressures in a shut in well, while the wells surrounding the test well are producing. This type of test provides information on reservoir connectivity, and preferential reservoir flow patterns, which can not be obtained from single well drawdown and buildup tests.

TRANSIENT FLUID FLOW THEORY

The mathematical basis for pressure analysis methods is the unsteady-state or transient fluid flow theory in a porous medium, which can be obtained by using (1) law of conservation of mass, (2) Darcy's law, and (3) equation of state.

The resultant radial flow for a slightly compressible fluid is given by the Diffusivity Equation[1,2,3]

$$\frac{\partial^2 p}{\partial r^2} + \frac{1}{r}\frac{\partial p}{\partial r} = \frac{1}{\eta}\frac{\partial p}{\partial t}$$

where:

$$\eta = \frac{k}{\phi \mu c}$$

p = pressure
r = radial direction of flow
t = time
k = permeability
ϕ = porosity
μ = fluid viscosity
c = fluid compressibility

The name comes from the similar equation, derived much earlier, which applies to the diffusion of heat. If the right hand side of this equation is zero, we have steady state conditions and this becomes Darcy's Equation.

CONVENTIONAL ANALYSIS

Earlier conventional techniques for analyzing well test data, including Horner, and Miller, Dyes and Hutchinson, are applicable for tests of sufficient duration, if and when radial flow is established in the reservoir and boundary effects are not too important.[4] These methods are essentially based upon solutions of the diffusivity equation.

The most common and useful of the many solutions of the diffusivity equation is called the constant terminal rate solution, for which the initial and boundary conditions are:

$p = p_i$ at $t = 0$ for all r
$p = p_i$ at $r = \infty$ for all t (infinite reservoir)
q = constant at $r = r_w$ for all t (line source)

An approximation of the constant terminal rate solution for long times and vanishing wellbore is:

$$p_w = p_i - \frac{162.6 q \mu B}{kh} \left[\log_{10} \frac{0.000264 kt}{\phi \mu c r_w^2} + 0.351 \right]$$

where:

p_w = wellbore pressure, psia
p_i = initial reservoir pressure, psia
q = flow rate, STB/D
μ = viscosity, cp
B = formation volume factor, RB/STB
k = permeability, md
h = zone thickness, ft
ϕ = porosity, fraction
r_w = wellbore radius, ft
c = compressibility, psia^{-1}
t = time, hr

This expression for the time-dependent well pressure due to constant rate production is the basic equation used in well pressure test analysis. According to this equation, a plot of wellbore pressure versus logarithmic (base 10) time should yield a straight line for a test, say drawdown, if the reservoir producing rate is constant.

Typical drawdown curve

A schematic representation of well pressure behavior in the case of drawdown is shown in Figure 9-2. It may be observed that the regions of the plot at early and late times are not straight but curved. Early time is affected by conditions at or close to the well. Middle time, which is straight or virtually straight, is influenced by the inter-well conditions. Late time is affected by interference from adjacent wells and/or reservoir boundaries.

The middle time slope can be used to determine formation flow capacity (permeability x thickness), kh. For example, slope, m, in the drawdown curve above is approximately 500 psi over 2 cycles. The flow capacity can be easily calculated as follows.

From Figure 9-2:

$$m = \frac{162.6q\mu B}{kh} = \frac{500psi}{2cycle} = 250psi\,/\,cycle$$

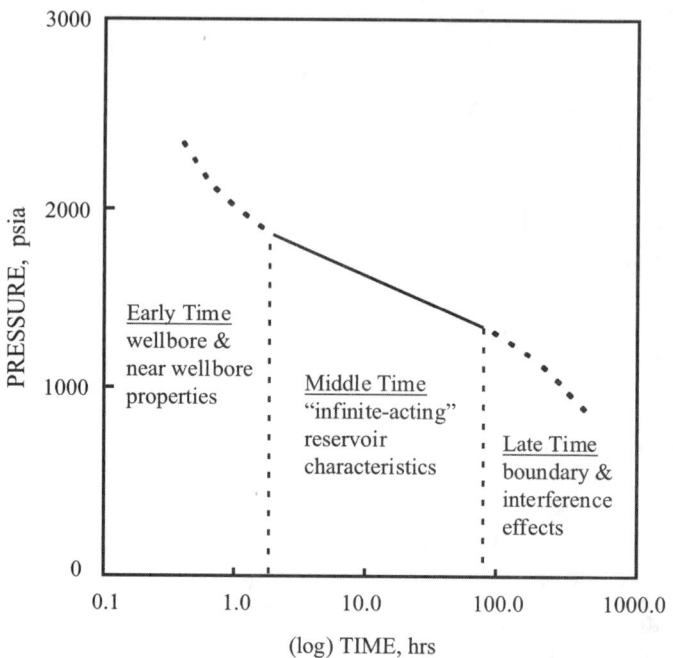

Fig. 9-2 • Drawdown Curve

if:

q = 153 STB/D
μ = 0.8 cp, and
B = 1.38 RB/STB

then:

$$kh = \frac{162.6x153x0.8x1.38}{250} = 109.9 \text{ mD-ft}$$

Buildup test analysis

The solution to the flow problem presented above is based upon constant flow rate. A powerful technique called the principle of superposition can be used to treat the more usual variable-rate case by adding together several constant-rate cases.

For example, suppose a well is produced at rate q for time t and then closed for a time interval Δt. The well pressure, by the principle of superposition and approximation of variable rate flow into the well, is given by the following equation:

$$p_w = p_i - 162.6\frac{q\mu B}{kh}\log_{10}\left(\frac{t+\Delta t}{\Delta t}\right)$$

This is the basic equation for pressure buildup analysis, based upon a solution for an infinite, homogeneous, one-well reservoir containing a fluid of small and constant compressibility.

It indicates that a plot of p_w versus log $(t+\Delta t)/\Delta t$ should yield a straight line. SPE Monograph 1[1] provides a better understanding and development of this equation, which is attributed to Horner.[5]

The theory and practice agree, except for wellbore damage due to skin, improvement due to fracturing, bounded reservoir, interference due to multiple wells, phase separation in tubing, fault, etc., as shown on Figure 9-3. The example buildup curves presented in this figure may be used for some qualitative interpretation of the actual curves by comparison.

Extrapolation of the straight line to $[(t + \Delta t)/\Delta t] = 1.0$ or $\log_{10} [(t + \Delta t)/\Delta t] = 0$ for infinite shut in time yields a pressure called p^*, the "extrapolated pressure" or Horner "false pressure." For a new well in an infinite reservoir, p^* is the initial reservoir pressure, p_i. If the well is produced for some time, p^* is approximately equal to the average pressure in the drainage area around the well.

TYPE-CURVE ANALYSIS

The type curve approach is very general, representing the pressure behavior of reservoirs with specific features, such as wellbore storage, skin, fractures, etc.[4] Type curves are usually log-log plots of dimensionless pressures versus dimensionless times. Each curve corresponds to some dimensionless parameter, characterizing the reservoir and wellbore conditions. The log scales enhance the visibility of curve variations.

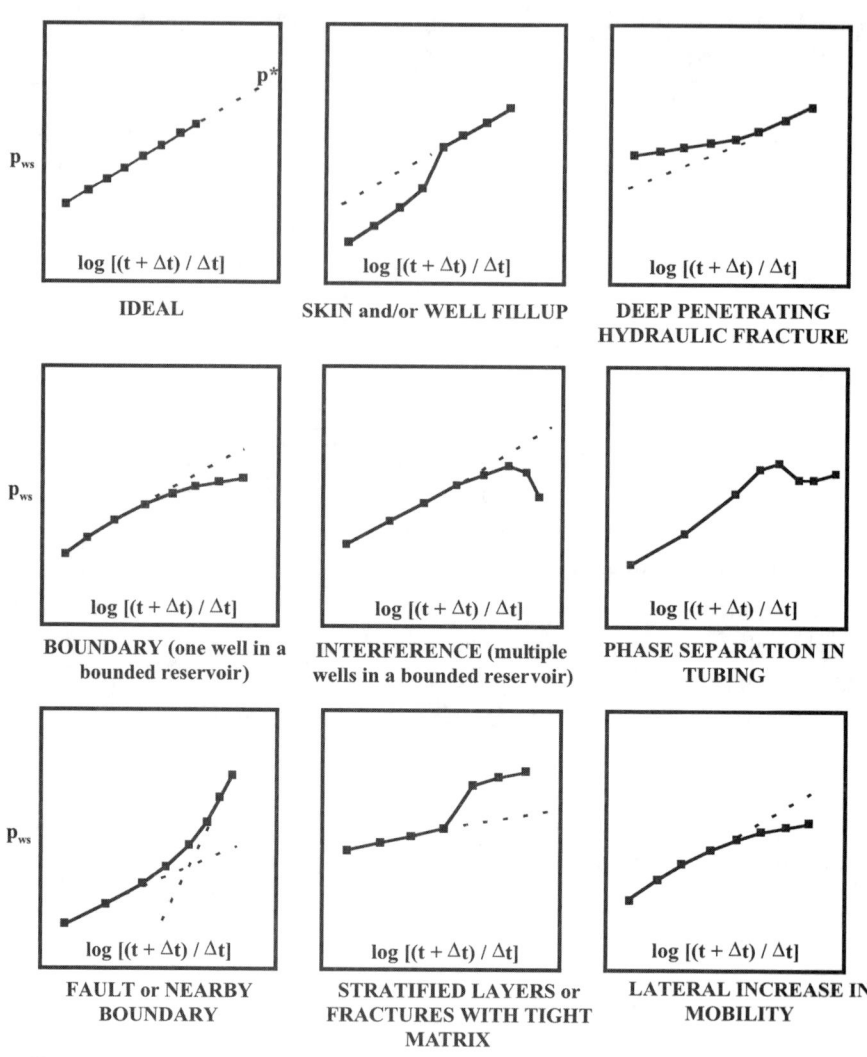

From Matthews & Russell [1]

Fig. 9-3 • Example Buildup Curves Illustrating Various Effects

For example, Figure 9-4 represents a type-curve for wellbore storage and skin effects for the case of a reservoir with infinite extent and homogeneous behavior. Dimensionless pressure, P_D, is plotted versus t_D / C_D, where each curve is characterized by $C_D e^{2S}$. This figure shows the limits of the various flow regimes, *i.e.*, end of storage and start of semi-log radial flow, and the ranges of various well conditions, *i.e.*, damaged, zero skin, acidized, and fractured.

Dimensionless parameters for wellbore storage and skin in a homogeneous infinite reservoir are:

Dimensionless Pressure,
$$P_D = \frac{kh}{141.2qB\mu}\Delta p$$

Dimensionless Time,
$$t_D = \frac{0.000264k\Delta t}{\phi\mu c_t r_w^2}$$

Dimensionless Wellbore Storage Coefficient,
$$C_D = \frac{0.8936C}{\phi c_t h r_w^2}$$

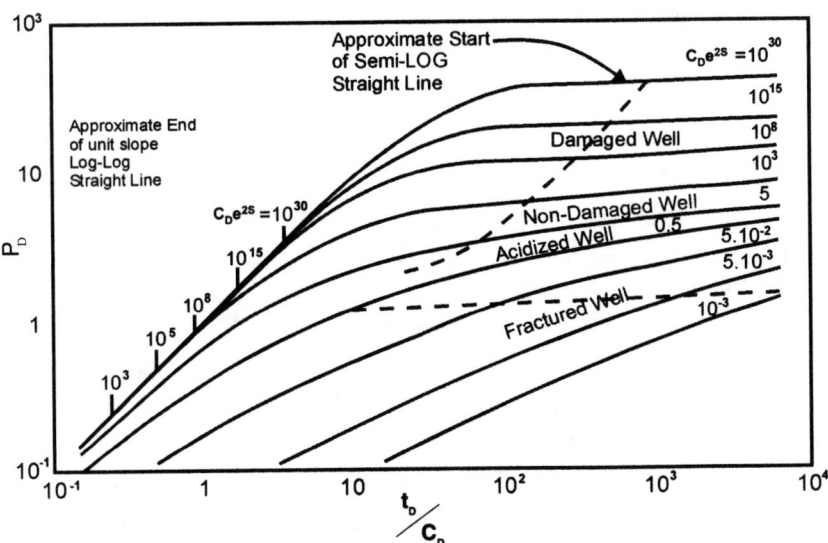

Fig. 9-4 • Type Curve for Wellbore Storage and Skin

Skin Factor, $\qquad\qquad\qquad\qquad\qquad$ $S = 0.5 \ln(C_D e^{2S} / C_D)$

The English field units of the variables are:

k, mD; h, ft; q, Bbl/D; μ, cp; Δp, psi; Δt, hrs; c_t
(total compressibility), 1/psi; r_w, ft; C
(wellbore storage coefficient), Bbl/psi.

Matching the field data to the appropriate type-curve yields:

• The permeability-thickness product, kh, from the pressure match
• The wellbore storage constant, C, from the time match
• The skin factor, S, from the curve match

In addition, the condition of the well can be determined from the $C_D e^{2S}$ value:

$C_D e^{2S} > 1000$	Damaged Well
$1000 > C_D e^{2S} > 5$	Normal Well
$5 > C_D e^{2S} > 0.5$	Acidized Well
$0.5 > C_D e^{2S}$	Fractured Well

Example problem

The interpretation method is illustrated with an example from a training manual of Scientific Software-Intercomp: "Well Test Interpretation in Practice," by Gringarten.[6] The well test consists of a 48-hour drawdown at a constant rate of 700 bbl/D, followed by a 72-hour buildup. Figure 9-5 presents pressure vs. time data, and reservoir parameters are given in Table 9-1.

The interpretation process starts with a log-log plot of build-up delta pressure vs. delta time data (Fig. 9-6). Our assumption is that the test well has wellbore storage and wellbore damage due to skin in a reservoir with homogeneous behavior. Therefore, buildup data need to be matched with the type curves of Figure 9-4, corresponding to $C_D e^{2S}$ greater than 1,000.

An initial type-curve match is shown in Figure 9-7, which indicates a

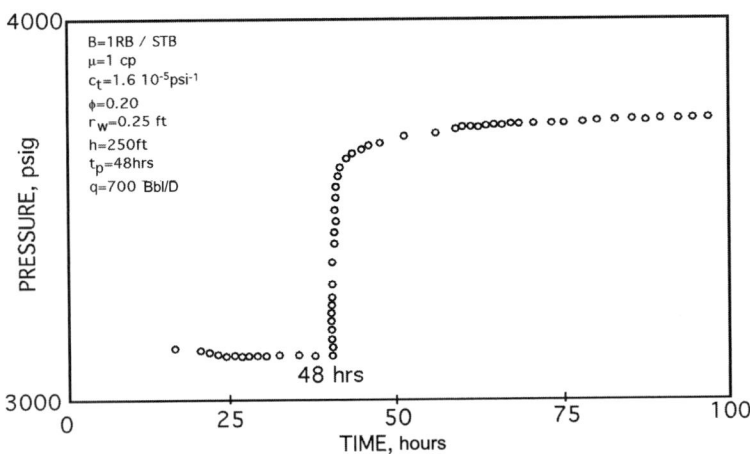

Fig. 9-5 • Pressure Data for Example Problem

Reservoir Thickness, feet 250
Reservoir Porosity, % 20
Reservoir Permeability, md 6
Total Compressibility, psi^{-1} . . . . 1.6×10^{-5}
Well Radius, foot 0.25
Reservoir Pressure ($\Delta t = 0$), psia 3140
Oil Formation Volume Factor, stb/rb 1.0
Oil Viscosity, cp 1.0
Production Time, hrs 48
Production Rate, stb/day 700
Buildup Time, hrs 72

Table 9-1 • Reservoir Data for Example Problem

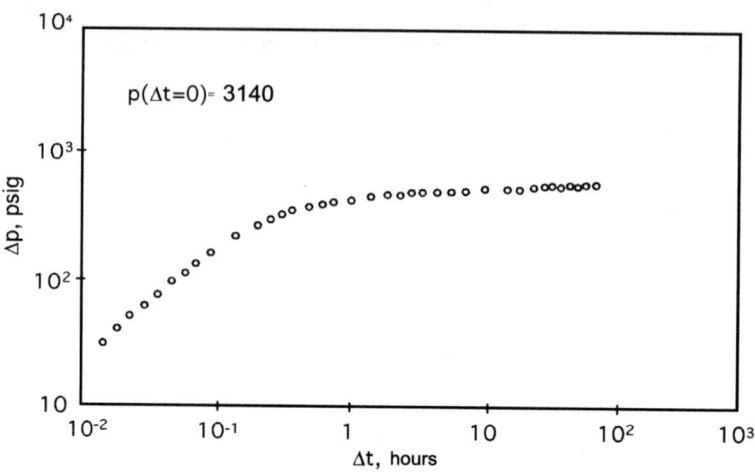

Fig. 9-6 • *Log-Log Plot of Build-up Pressure for Example Problem*

reasonable match to $C_D e^{2S} = 10^4$, with Horner analysis valid after four hours (pointer number 2). The purpose of this initial match is to confirm the diagnostic of the model and to check if Horner analysis is applicable. It can be seen that the buildup data do not match the drawdown type-curve after approximately 10 hours (pointer number 3). This indicates that the semi-log Horner plot is only valid between 4 and 10 hours. Furthermore, the match

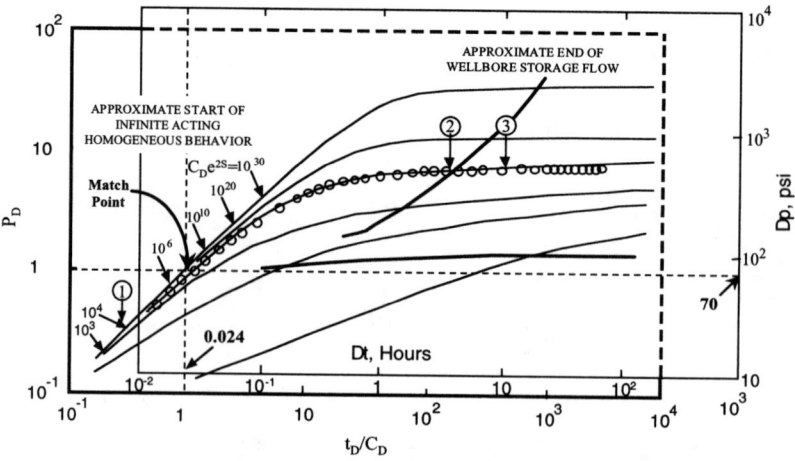

Fig. 9-7 • *Initial Type-Curve Match for Example Problem*

indicates the wellbore storage specialized analysis is only applicable to measurements taken before a Δt of 30 seconds (pointer number 1). Calculations for kh, C, and S are shown in Table 9-2, using Figure 9-6, log Δp versus log Δt curve, and Figure 9-7, log P_D versus log t_D/C_D curve.

A Horner plot is shown in Figure 9-8. The Horner straight line starts at 4 hours as predicted from the type-curve match. It has a slope of 83.2 psi/log cycle. At $[(t_p + \Delta t)/\Delta t] = 1$, $p^* = 3756.7$ psi. Calculations for flow capacity

---

Match results

$$PM = \frac{(P_D)_{match}}{(\Delta p)_{match}} = \frac{1}{70} = 0.0143$$

$$TM = \frac{(t_D/C_D)_{match}}{(\Delta t)_{match}} = \frac{1}{0.024} = 41.7$$

$$(C_D e^{2S})_{match} = 10^4$$

Analysis results $kh = 141.2\, qB\mu\, PM = 141.2\,(700)\,(1)\,(1)\,(.0143) = 1413$ md-ft

$$C = 0.000295 \frac{kh}{\mu} \frac{1}{TM} = 0.000295 \frac{1413}{1} \frac{1}{41.7} = 0.01\,\text{bbl/psi}$$

$$C_D = \frac{0.8936C}{\phi c_t hr_w^2} = \frac{0.8936(0.01)}{(0.2)(1.6x10^{-5})(250)(0.25)^2} = 178.7$$

$$S = 0.5\ln\left[\frac{(C_D e^{2S})_{match}}{C_D}\right] = 0.5\ln\left[\frac{10^4}{178.7}\right] = 2.0$$

$$\text{Horner slope, } m = \frac{1.151}{PM} = \frac{1.151}{0.0143} = 80.5 \text{ psi/log cycle}$$

Table 9-2 • Log-Log Type—Curve Calculations for Example Problem

(kh), effective permeability (k), skin factor (S), pressure match, PM = $(p_D)_{match}/(\Delta p)_{match}$, are shown in Table 9-3.

A comparison of the results of the Horner and Log-Log plot analysis methods is shown in Table 9-4.

QUESTIONS TO BE ASKED BEFORE STARTING A WELL TEST ANALYSIS

Gringarten[6] recommends the following questions to be asked before starting a well test analysis:

Reservoir questions

1. What type of rock?
   - If limestone, carbonate, granite, basalt, or loose sand, expect double porosity behavior
   - If acidized carbonate (radius of investigation, $r_i \cong 3$ ft), expect composite behavior
   - If consolidated sandstone, do not expect double porosity behavior

2. Is this a layered reservoir?

3. Any known boundary, producing or injecting well nearby, gas cap, and/or water contact?

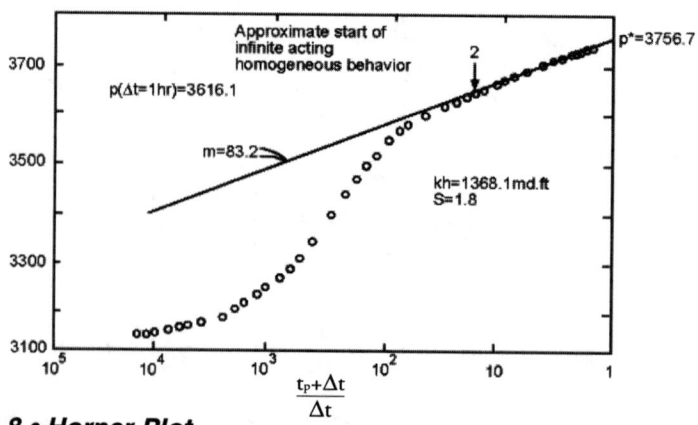

Fig. 9-8 • Horner Plot

From Figure 9-8,
  m, psi/log cycle = 83.2
  p*, psi = 3756.7
  p ($\Delta t = 1$ hr) on Horner straight line, psi = 3616.1

Analysis results

$$kh = 162.6 \frac{qB\mu}{m} = 162.6 \frac{(700)(1)(1)}{83.2} = 1368.0 \, \text{md-ft}$$

$$k = \frac{1368.0}{250} = 5.47$$

$$S = 1.151 \left\{ \frac{p(\Delta t = 1hr) - p(\Delta t = 0)}{m} - \log\left[\frac{k}{\phi\mu c_t r_w^2}\right] + \log\left[\frac{t_p + 1}{t_p}\right] + 3.23 \right\}$$

$$S = 1.151 \left\{ \frac{3616.1 - 3140}{83.2} - \log\left[\frac{5.47}{(0.2)(1)(1.6x10^{-5})(0.25)^2}\right] + \log\left[\frac{48 + 1}{48}\right] + 3.23 \right\}$$

$$S = 1.151 \, (5.722 - 7.437 + 0.009 + 3.23) = 1.76$$

$$PM = \frac{1.151}{m} = \frac{1.151}{83.2} = 0.0138$$

Table 9-3 • Horner Calculations for Example Problem

Analysis Method	kh	C	S	p
Log-Log Type-Curve	1413	0.01	2.0	
Horner	1368		1.76	3756.7

Table 9-4 • Check of Interpretation Consistency

Well questions

4. Is the well vertical or horizontal?

5. What was done to the well?
   - acidized
   - fractured

6. Has the well been drilled through the entire formation?

7. How long has the well been in production?

Fluid questions

8. What is the dominant phase?
   - oil, gas, or water

9. How many phases?
   - in the wellbore
   - in the reservoir

10. What is the bubble point pressure (oil) or dew point pressure (condensate gas)?
    - expect composite behavior in low permeability reservoirs ($r_i \cong 50$ ft) if pressure falls below bubble point (oil) or dew point (condensate gas)

CHECKS DURING ANALYSIS

The following checks to be made during the analysis are also from Gringarten:

Data validation

1. Check pressure gauges
2. Check start and end of flow periods
3. Check rate consistency

Model diagnostic

4. Check time and pressure at start of flow period

5. Check derivative smoothing

6. Remember your 10 questions

7. Use common sense

Matching and final results

8. Check $(P_{av})_i$ on the simulation
   If different from Horner Match, keep simulation $(P_{av})_i$ and regress on the other parameters on Horner Match

9. Check result consistency between flow periods

10. Use common sense

WORKBENCH WELL TEST ANALYSIS SOFTWARE

The well test analysis module of Baker-Hughes/SSI's Petroleum WorkBench combines type-curve analysis with conventional semi-log analysis and can be used to interpret data in homogeneous or heterogeneous formations.[7] It contains a variety of near-wellbore, reservoir, and boundary options.

To perform a well test analysis, at a minimum, the following data are needed:

- Pressure history data
- Production rates for each flow period
- Well and reservoir parameters

For test design, the following data are needed:

- The model type
- Rates for multiple flow periods
- Estimates of parameter values
- The initial pressure value

Pressure, temperature, and history rate data can be loaded from ASCII files in any format. Also multi-gauge data can be imported and used for a single analysis. As many as 50,000 data points can be loaded for each gauge.

Well test analysis

Once the data are specified, WorkBench automatically creates an appropriate type-curve, based on user-defined models for near-wellbore, reservoir behavior, and boundary effects:

Early times—near wellbore effects:

- Wellbore storage and skin
- Line source solution (for interference tests)
- Infinite conductivity or uniform flux vertical fracture
- Finite conductivity vertical fracture

Middle times—reservoir behavior:

- Homogeneous
- Double porosity
- Double permeability
- Composite reservoir
- Composite, double porosity

Late times—boundary effects:

- Infinite lateral extent
- Single boundary, channel boundaries, open-ended rectangle, rectangle
- Partially communicating boundary (leaky fault)
- Intersecting boundaries (wedge)

The interpretation process is always started by selecting a flow period to analyze and identifying the interpretation model that applies to the data. There are two diagnostic options available for identifying the interpretation model. They are:

- Log-log diagnostic analysis
- Horner analysis

Each of these consists of a menu of options within the software that automatically builds a model and generates the corresponding type curve for matching to the data.

A regression algorithm is available to obtain the best match from the data. The user can select which variables are to be used in the regression. In addition, the original data can be either time or pressure shifted to obtain a better match.

The log-log and dimensionless Horner matches ensure that the analysis is consistent for the flow period of interest. It is necessary, however, to verify that the analysis is also valid for the entire pressure history. This is done by simulating the pressure history using the actual rate data, the interpretation model, the numerical values of the well, and reservoir parameters obtained from the analysis. The computed pressure history can be easily compared to the measured data. They must match for the analysis to be consistent.

Well test software makes it quite easy to refine the analysis until differences are resolved and a best fit between the data and a model is obtained. Then a simulated pressure history can be calculated and compared to the recorded history. Figure 9-9 shows solutions of the previously discussed example problem. A first simulation using results from Horner analysis gives too high a pressure drop during the drawdown, indicating too low a *kh* (1,368 md-ft), although the match during the buildup is good. A simulation

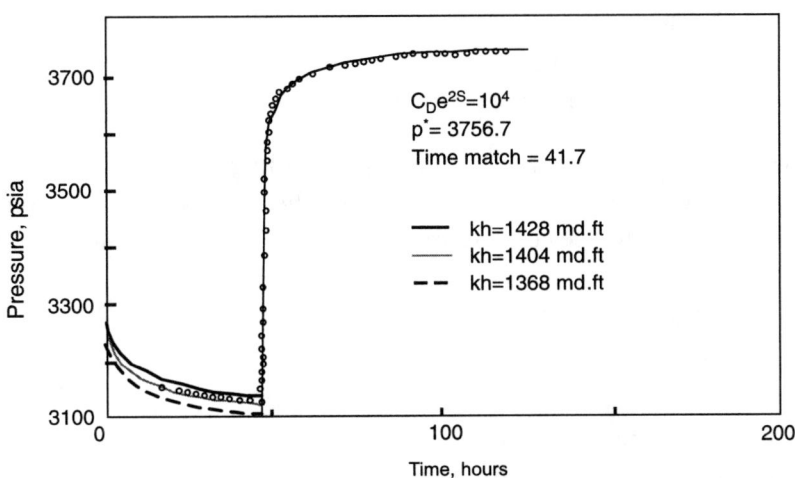

$C_D e^{2S} = 10^4$
$p^* = 3756.7$
Time match = 41.7

— kh=1428 md.ft
— kh=1404 md.ft
– – kh=1368 md.ft

Fig. 9-9 • Simulated Pressure Histories for Example Problem

with *kh* = 1,428 md-ft yields too low a pressure drop during drawdown, but still matches during buildup. A final *kh* value of 1,404 md-ft gives a good fit to the entire test.

Well test design

Design features of the software can be used to generate the expected rate-normalized pressure response for constant-rate production. The well and reservoir data, selection of a model, and its corresponding parameter values are needed. Based on the rate-normalized response, the rate sequence can be defined and the goals of the design can be reached. Users can display the log-log and superposition plots for the various flow periods to help them determine whether the design is adequate. All the plots show both type-curves and the expected measured data points that will provide the best definition for the test.

A well-conducted test will enable users to reach reasonably confident conclusions concerning well and reservoir features when the test data are analyzed. In particular, the analysis of the test quantifies the effects of:

- well damage
- hydraulic fracture length
- permeability
- double-porosity behavior
- distance to boundaries

With a poorly designed test, it may be difficult or impossible to analyze certain reservoir characteristics. For example, the test may be long enough to show wellbore storage characteristics, but far too short to show distances to boundaries.

There are several factors that effect the test quality:

- Poor data recording or sampling
- Inadequate rate information
- Insufficient test duration
- Infrequent sampling at early times, which may mask effects of hydraulic fractures and other characteristics

The test design can clearly show:

- whether the sampling is frequent enough to show early time behavior
- how long the test should be within a range of estimated parameter values

REFERENCES

1. Mathews, C. S., and Russell, D. G.: *Pressure Buildup and Flow Tests in Wells*, Monograph Volume 1, SPE, Dallas (1967)

2. Earlougher, R.C.: Advances in Well Test Analysis, Monograph Volume 5, SPE, Dallas (1977)

3. Lee, John: *Well Testing*, Society of Petroleum Engineers of AIME, Dallas (1982)

4. Gringarten, A. C., Bourdet, D. P., Landet, P. A., Kniazeff, V. J.: "A Comparison Between Different Skin and Wellbore Storage Type Curves For Early Time Transient Analysis." SPE 8205, paper presented at the SPE Annual Fall Technical Conference and Exhibition, Dallas, Texas, Sept. 23-26, 1979

5. Horner, D. R.: "Pressure Buildup in Wells," Proceedings., Third World Pet. Congress., E. J. Brill, Leiden (1951) II, 503

6. Gringarten, A.: *Well Test Interpretation Practice*, Scientific Software-Intercomp Inc., Denver, CO

7. Petroleum WorkBench: *Well Test Analysis - Interpret/2 User Manual*, Scientific Software-Intercomp, Inc., Denver, CO, November 21, 1954

Chapter *TEN*

PRODUCTION PERFORMANCE ANALYSIS

The evaluation of the past and present performance of a reservoir and the forecast of its future behavior is an essential aspect of reservoir management. Satter and Thakur presented the techniques used for reservoir performance analyses.[1] Users of production analysis software need to understand the concepts, theory, and limitations behind the techniques in order to apply them effectively.

OIL AND GAS RESERVOIRS

Figure 10-1 depicts an example of hydrocarbon phase behavior, *i.e.*, pressure vs. temperature for a given fluid composition. Reservoir type, whether oil or gas, depends upon the location of the point representing the reservoir pressure and

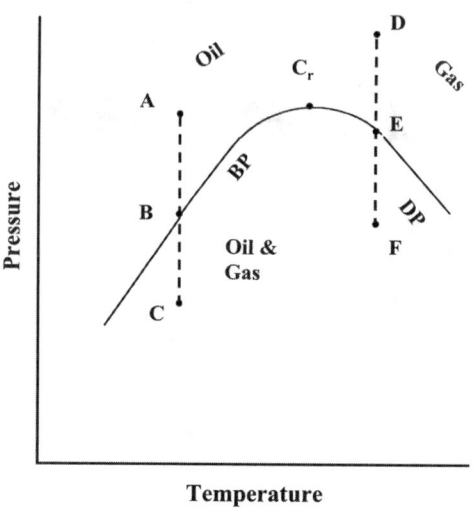

Fig. 10-1 • Hydrocarbon Phase Behavior

temperature at the time of discovery. For example, points A, B, and C represent under-saturated oil, bubble point, and saturated oil and gas, respectively. Points D, E, and F represent dry gas, dew point, gas and condensate liquid. Figure 10-2 shows under-saturated, and saturated oil reservoirs, and also a gas reservoir in contact with an aquifer.

NATURAL PRODUCING MECHANISMS

Primary performance of oil and gas reservoirs is dictated by natural viscosity, gravity, and capillary forces. Factors influencing the reservoir performance are geological characteristics, rock and fluid properties, fluid flow mechanisms, and production facilities. The natural producing mechanisms influencing the primary reservoir performance are shown in Figure 10-3. As can be noted, recovery is dramatically influenced by the type of drive mechanism.

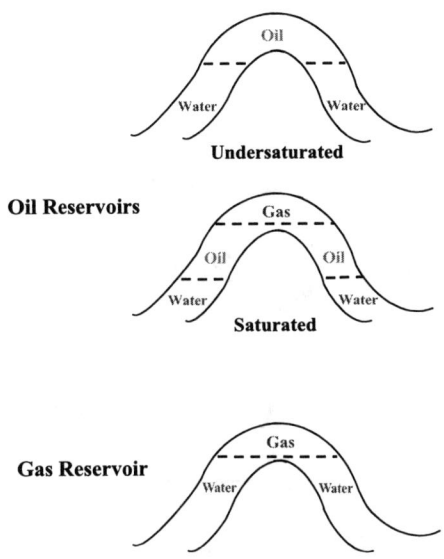

Fig. 10-2 • Hydrocarbon Reservoirs

RESERVOIR PERFORMANCE ANALYSIS

Commonly used reservoir performance analysis/reserves evaluation techniques and the estimates they provide are:

- Volumetric
    - original hydrocarbon in place
- Decline curve
    - reserves
    - ultimate recovery
- Classical material balance
    - original hydrocarbon in place
    - recovery mechanism
- Mathematical simulation

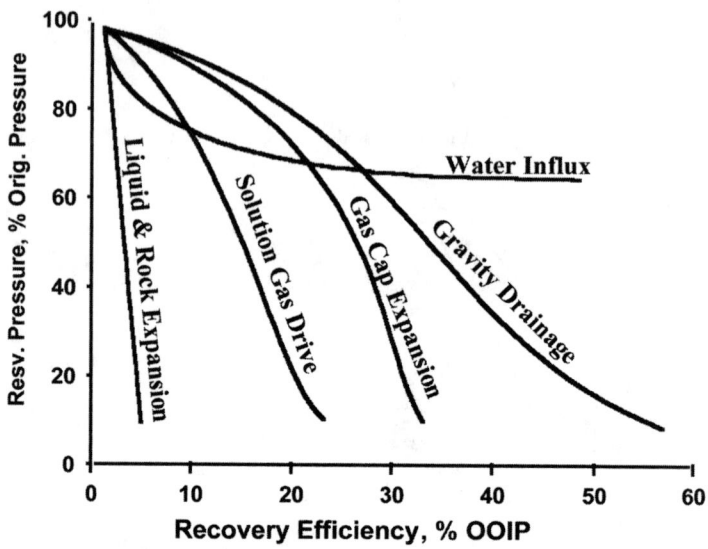

Fig. 10-3 • Effects of Reservoir Producing Mechanisms on Recovery Efficiency

- original hydrocarbon in place
- reserves
- ultimate recovery
- performance under various scenarios

These techniques will be discussed in the following chapters. The accuracy of the performance analysis is dictated by the depth of understanding of the reservoir characteristics, *i.e.*, reservoir model, and the quality of the techniques used. Table 10-1 presents applicability, accuracy, data requirements, and results of the various techniques.

RESERVES

Hydrocarbon reserve is defined as the future economically-recoverable hydrocarbon from a reservoir. Reserve can be defined as (Fig. 10-4):

Reserve = Ultimate Economic Recovery – Past Cumulative Production

	Volumetric	Decline Curve	Material Balance	Mathematical Models
Applicability/ Accuracy				
Exploration	Yes / ?	No	Yes / ?	Yes /?
Discovery	Yes / ?	No	Yes / ?	Yes / ?
Delineation	Yes / ?	No	Yes / ?	Yes / Fair
Development	Yes / Fair	No	Yes / Fair	Yes / Good
Production	Yes / Fair	Yes / Fair	Yes / Good	Yes / Very Good
Data Requirements				
Geometry	A,h	No	A,h	A,h
Rock	Φ,S	No	Φ,S, K_r,C	Φ,S, K_r,C
Fluid	B	No	PVT homogeneous	VT heterogeneous
Well	No	No	PI for rate vs. time	locations, perforations, PI
Production & Injection	No	Production	Yes	Yes
Pressure	No	No	Yes	Yes
Results				
Orig. Hydrocarbon in Place	Yes	No	Yes	Yes
Ultimate Recovery	Yes with rec. eff.	Yes	Yes	Yes
Rate vs. Time	No	Yes	Yes with PI	Yes
Pressure vs. Time	No	No	Yes with PI	Yes

Table 10-1 • Comparison of Reservoir Performance Analysis/Reserves Estimation Techniques

Reserve is attributed to primary, secondary, or tertiary recovery processes. It is changing due to additional production and also due to any revision in ultimate production.

Ultimate recovery is given by:

Ultimate Recovery = Original Hydrocarbon In Place x Recovery Efficiency

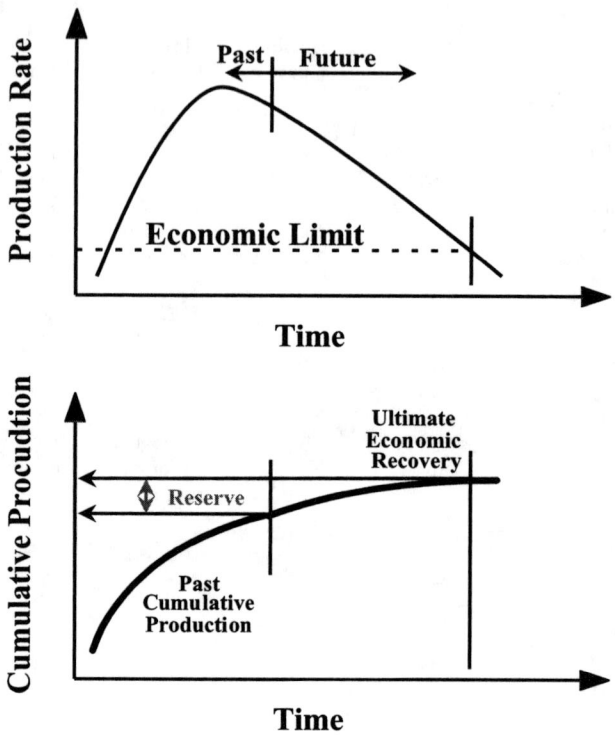

Reserve = Future Production =
Ultimate Economic Recovery - Past Cumulative Production

Fig. 10-4 • Definition of Reserve

In addition to the factors discussed earlier, recovery efficiency will depend upon the primary, secondary, or tertiary method of producing the reservoir. The quality of reservoir operation/management is also very important.

REFERENCES

1. Satter, A., and Thakur, G. C.: *Integrated Petroleum Reservoir Management: A Team Approach,* PennWell Books, Tulsa, Oklahoma (1994)

Chapter *ELEVEN*

VOLUMETRIC METHOD

INTRODUCTION

Estimates of oil and gas reserves fall into three main categories: analogy, volumetric, and performance. The method of analogy is based upon historical experience from similar reservoirs and wells in the same area and with similar depositional environment. It is particularly useful when limited data are available for the reservoir. The volumetric method for reserve determination applies after several wells have been drilled and found to contain hydrocarbons. The performance methods involve general material balance, decline curve analysis, and numerical simulation studies, which are usually done after gathering data for an extended period of production.

The volumetric method of computing original oil or gas in place in a reservoir is usually carried out as soon as sufficient geological data is available. This method is one of the most fundamental and elementary tools in assessing the value of a reservoir. It is based upon integration of geological, geophysical, petrophysical, and reservoir data as listed below:

- Field and lease maps
- Structure and isopach maps
- Open-hole and cased-hole logs
- Core analysis
- Porosity and permeability, with cut-offs
- Elevations of oil/water and gas/oil contacts
- Water saturations, with cut-offs
- Basic fluid properties, such as formation volume factor and solution gas-oil-ratio

The volumetric calculations provide a reality check on more sophisticated classical material balance and reservoir simulation results.

RESERVOIR VOLUME

There are two ways of computing basic reservoir volumes. The first method uses geological structure maps, contoured on the subsea depths of both the top and the base of the reservoir zone in question. The area encompassed by each contour is planimetered and plotted on a graph of subsea depth versus area in acres, as shown in Figure 11-1. Gas-oil and oil-water contacts are also noted on the plot. The gross volumes of the gas- and oil-bearing zones may be determined by integrating the areas under the curves and accounting for the fluid contacts.

The more frequently used second method involves creating separate isopach maps for the net oil and/or net gas pay zones, using individual well-log data. These net sand isopach maps are developed when sufficient numbers of wells with data are available for analysis. A general rule of thumb would be 5-10 wells (nicely dispersed) as a minimum number for making a reliable map. The more wells are available, the better the isopach

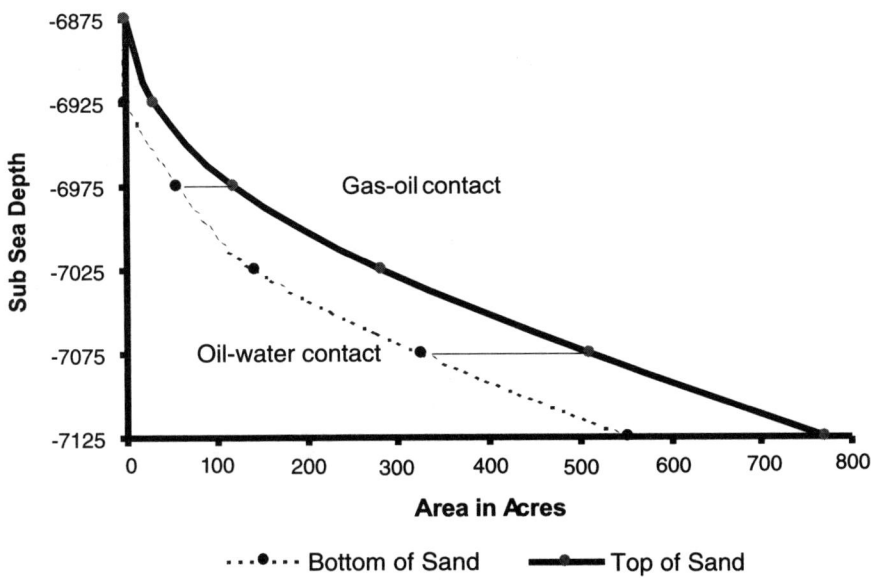

Fig. 11-1 • Subsea Depth Determining Reservoir Volume

map will reflect the actual reservoir character. Figure 11-2 is an example of a net pay isopach.

The determination of net pay is complicated by a number of factors:

- A directionally drilled well needs to have log measured depths converted to true vertical thickness, which accounts for the well deviation as well as the dip and strike of the formation
- Location of fluid contacts could differ from well to well as a result of a varying transition zone
- There must be sufficient effective porosity for commercial production; in some cases, a porosity cut-off is needed for calculations
- Formation permeability must be such that production rates of oil and/or gas are commercially viable

Once the isopach maps have been developed using evenly spaced contour increments, the next step is to compute the available reservoir volume

Fig. 11-2 • Example Net Pay Isopach

in acre-feet. This is usually done by planimetering the area enclosed by each contour in acres. An example is shown in Figure 11-3, which is a plot of contour area versus contour interval. The volume of the reservoir, V_o, may be found by integrating the area under the curve:

$$V_0 = \int_0^H f(h)dh \qquad (11\text{-}1)$$

where:
H = total reservoir thickness
$f(h)$ = equation of the curve in Figure 11-3
dh = equals the contour increment

An analytical expression for the area under the curve would be difficult to develop. However, this area can be determined by graphical integration. There are two primary means for determination as given below:

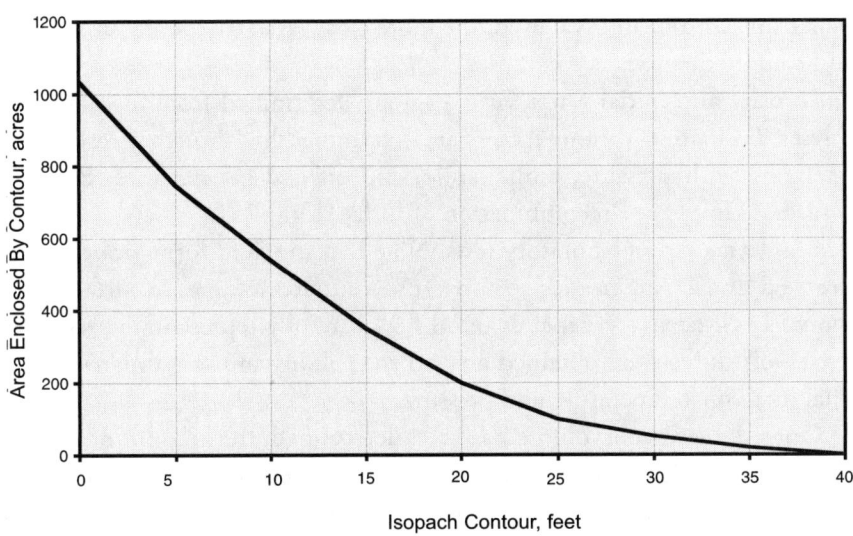

Fig. 11-3 • Plot of Contour Area vs. Contour Interval

Trapezoidal Rule

$$V = \tfrac{1}{2}H \, (a_0 + 2a_1 + 2a_3 + ... + 2a_{n-1} + a_n) + t_n a_n \qquad (11\text{-}2)$$

Simpson's Rule

$$V = \tfrac{1}{3} H \, (a_0 + 4a_1 + 2a_2 + 4a_3 + ... + 2a_{n-2} + 4a_{n-1} + a_n) + t_n a_n \quad (11\text{-}3)$$

where:

V = reservoir volume in acre-feet

a_0 = area enclosed by the zero contour (at oil-water or gas-oil contact), acres

a_1 = area enclosed by the first contour, acres

a_2 = area enclosed by the second contour, acres

a_n = area enclosed by the nth contour, acres

t_n = average formation thickness above the top contour, feet

Although Simpson's Rule is a little more tedious and requires an equal number of contour intervals, it tends to be the more accurate method. The smaller the contour interval used, the more closely the true reservoir volume will be approximated.

Porosity and initial water saturation are determined from log and core analyses. Formation volume factors are determined by laboratory tests or by correlation with pressure, temperature, and oil and gas gravities. Either a subsurface sample or a recombination of surface samples from separators and stock tanks is used in laboratory tests. Note that the fluid formation volume factors (B_o and B_g) are needed to convert sub-surface volumes to surface conditions. Their accuracy depends upon how carefully representative samples of reservoir fluids were obtained and on the validity of the estimates of the initial reservoir temperature and pressure.

Once the reservoir volume has been determined, the amount of oil and gas initially in place can be calculated from the relationships given below, using average rock and fluid properties.

Free gas or gas cap

With no residual oil present in the gas cap,

$$G = \frac{43,560 V \phi (1 - S_{wi})}{B_g} \qquad (11\text{-}4)$$

where:
G = in-place gas, standard cubic feet
V = reservoir volume, acre-feet
ϕ = reservoir porosity, faction
S_{wi} = connate water saturation
B_g = gas formation volume factor, rb/scf
43,560 = cubic feet/acre-foot

Oil in place

With no free gas present in the oil,

$$N = \frac{7758 V \phi (1 - S_{wi})}{B_o} \qquad (11\text{-}5)$$

where:

N = oil in place, stock-tank barrels

V = reservoir volume, acre-feet

ϕ = reservoir porosity, fraction

S_{wi} = connate water saturation, fraction

B_o = oil formation volume factor, rb/stb

7,758 = barrels/acre-foot

Solution gas in oil reservoir

$$G_s = R_s N \qquad\qquad (11\text{-}6)$$

where:

G_s = solution gas in place, standard cubic feet

R_s = solution gas-oil-ratio, scf/stb

To account for variations in porosity and saturations in the reservoir, contour maps of thickness x porosity x initial oil saturation can be prepared. This will give more accurate estimates of original hydrocarbon in place.

Ultimate recovery can be estimated by multiplying the OOIP or OGIP by the recovery efficiency factor. The oil recovery efficiency factor (STB/acre-ft or % OOIP) and the gas recovery factor (MMSCF/acre-ft or % OGIP) can be estimated from the performance data from similar or offset reservoirs. Published API correlations can be used to estimate primary recovery efficiency for oil reservoirs.[1,2,3] For gas reservoirs without the aid of water influx, the recovery efficiency is usually high, (80-90% OGIP). Recovery from active water drive reservoirs can be substantially less, (50-60% OGIP), which is attributed to high abandonment pressure, gas entrapment, and water coning.

REFERENCES

1. Arps, J. J., et al.: "A Statistical Study of Recovery Efficiency,"
   API Bulletin D14 (1967)
2. Doscher, T. M., et al.: "Statistical Analysis of Crude Oil Recovery
   and Recovery Efficiency," API Bulletin D14 (1984)
3. Satter, A. and Thakur, G. C.: *Integrated Petroleum Reservoir Management:
   A Team Approach*, PennWell Books, Tulsa, OK (1994)

Chapter *Twelve*

Decline Curve Method

When sufficient production data are available and production is declining, the past production curves of individual wells, lease or field can be extended to estimate future performance (Fig. 12-1). There are two very important assumptions in using decline curve analysis:

- Sufficient production performance data are available and decline in rate has been established
- All factors that influenced the curve in the past remain effective through out the producing life

Many factors, such as proration, changes in production methods, workovers, well treatments, pipeline disruptions, and weather and market conditions, influence production rates and, consequently, decline curves. Therefore, care

- Predict Future Production Rates
- Forecast Reserves

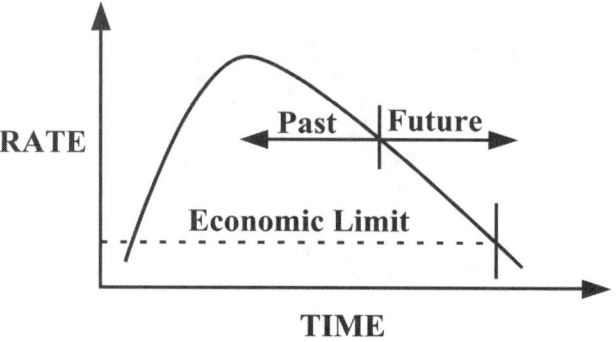

Fig. 12-1 • Decline Curve Predictions

must be taken in extrapolating the production curves into the future. When the shape of a decline curve changes, the cause should be determined, and its effect upon the reserves evaluated.

DECLINE CURVES

The commonly used decline curves for oil reservoirs are:

- Log Production Rate vs. Time (Fig. 12-2)
- Production Rate vs. Cumulative Production (Fig. 12-2)
- Log of Water Cut vs. Cumulative Production (Fig. 12-3)
- Oil-Water or Gas-Oil Contact vs. Cumulative Production
- Log Cumulative Gas Production vs. Cumulative Oil Production

When production rate plots (Fig. 12-2) are straight lines, they are called constant rate or exponential decline curves. Since a straight line can be easily extrapolated, exponential decline curves are most commonly used. In case of harmonic or hyperbolic rate decline, the plots show curvature (Fig. 12-2). Both the exponential and harmonic decline curves are special cases of hyperbolic decline curves. Unrestricted early production from a well shows hyperbolic decline rate. However, constant or exponential decline rate may be reached at a later stage of production.

Water cut curves (Fig.12-3) are employed when economic production

Log of Production Rate vs. Time

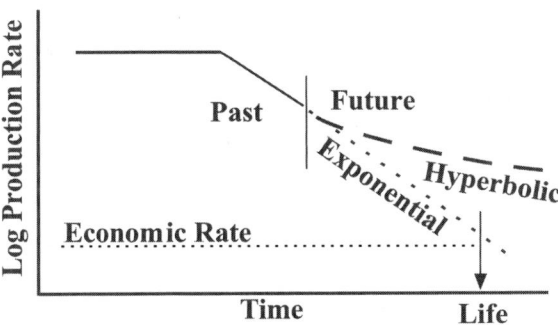

Production Rate vs. Cumulative Production

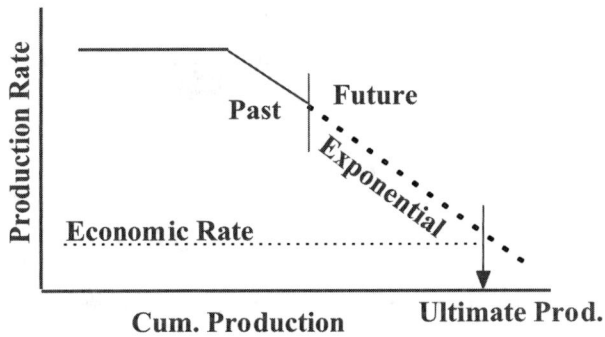

Fig. 12-2 • Production Rate Plots

131

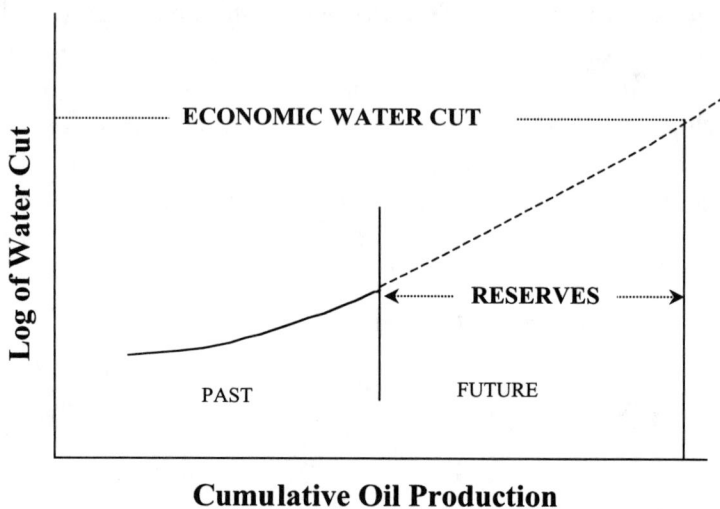

Fig. 12-3 • Log of Water Cut vs. Cumulative Oil Production

rate is dictated by the cost of water disposal. A straight-line extrapolation of log of water cut vs. cumulative oil production may not be reasonably done in the higher water cut levels. It may yield a conservative estimate of reserves. On the other hand, if oil cut data is used instead of water cut in the same levels, straight-line extrapolation of log of oil cut vs. cumulative oil production may deteriorate and lead to overly optimistic reserve estimates.

For gas reservoirs, a *p/z* vs. Cumulative Production plot gives a straight line for depletion drive reservoirs, where *p/z* is the average reservoir pressure (*p*) divided by gas compressibility factore (*z*) at that pressure.

DECLINE CURVE EQUATIONS

A general mathematical expression for the rate of decline, D, is shown in Figure 12-4, where *K* and *n* are constants. The type of decline is determined by the value of *n*. Standard curve types are:

Type of Decline	n	D
Exponential	0	constant
Hyperbolic	0<n<1	variable
Harmonic	1	variable

$$D = -\frac{dq/dt}{q} = Kq^n$$

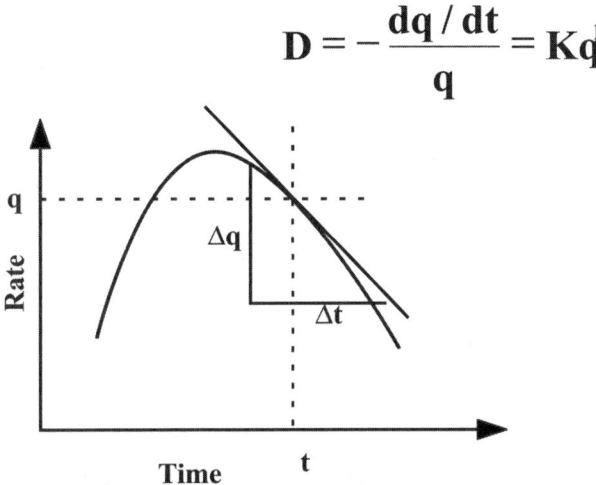

Fig. 12-4 • Decline Curve Methods

- *Exponential,* where decline is a constant percentage
- *Harmonic,* where decline is proportional to the rate
- *Hyperbolic,* where decline is proportional to a fractional power, *n,* of the rate; depending on the *n* value, the hyperbolic equation can model harmonic or exponential decline or something in between

Equations for production rates (*q*) and cumulative production (*Q*) for the various types of decline curves[1,2] are shown below:

133

A general mathematical expression for the rate of decline, D can be expressed as:

$$D = - \frac{dq/dt}{q} = Kq^n \qquad (12\text{-}1)$$

where:

q = production rate, barrels per day, month, or year
t = time, day, month, or year
K = constant
n = exponent

The decline rate in Equation (12-1) can be constant or variable with time, yielding three basic types of production decline as follows:

1. *Exponential or constant decline*

$$D = - \frac{dq/dt}{q} = K = - \frac{\ln\left(\frac{q_t}{q_i}\right)}{t} \qquad (12\text{-}2)$$

where:

$n = 0$, K = constant
q_i = initial production rate
q_t = production rate at time t

The rate-time and rate cumulative relationships are given by:

$$q_t = q_i \cdot e^{-Dt} \qquad (12\text{-}3)$$

$$Q_t = \frac{q_i - q_t}{D} \qquad (12\text{-}4)$$

where:

Q_t = cumulative production at time t

$$(12\text{-}5)$$

A familiar decline rate for exponential decline is as follows:

$$D' = \frac{\Delta q}{q_i}$$

where:

Δq is the rate change in the first year
In this case the relationship between D and D' is given by:

$$D = -ln\left(1 - \frac{\Delta q}{q_i}\right) = -ln\left(1 - D'\right)$$

(12-6)

2. Hyperbolic decline

$$D = -\frac{dq/dt}{q} = Kq^n \quad (0 < n < 1)$$

(12-7)

Note that this is the same as the general decline rate equation (12-1), except for the constraint on n.

For initial condition

$$K = \frac{D_i}{q_i^n}$$

the rate-time and rate-cumulative relationships are given by:

$$q_t = q_i(1 + n D_i t)^{-1/n}$$

(12-8)

$$Q_t = \frac{q_i^n\left(q_i^{1-n} - q_t^{1-n}\right)}{(1 - n) D_i}$$

(12-9)

where:

D_i = initial decline rate

3. Harmonic decline

$$D = -\frac{dq/dt}{q} = Kq$$

(12-10)

where:

$n = 1$

For initial condition:

$$K = \frac{D_i}{q_i}$$

The rate-time and rate-cumulative relationships are given by:

$$q_t = \frac{q_i}{(1 + D_i t)}$$

(12-11)

$$Q_t = \frac{q_i}{D_i} \ln \frac{q_i}{q_t}$$

(12-12)

135

ANALYSIS PROCEDURE

The general procedure for analysis is as follows:
- Gather available production performance data
- Observe plots of log of oil rate vs. time, and log of water cut vs. cumulative production
- Determine reasons for production data anomalies (upward or downward)
- Evaluate various means of plotting the data
- Select the portion of the performance data applicable for analysis
- Perform regression for history match
- Forecast future performance with economic oil production rate and water cut

EXAMPLE

Standard method

An example is presented to illustrate application of decline curve analysis using Baker-Hughes/SSI's Production Data Analysis program. Figure 12-

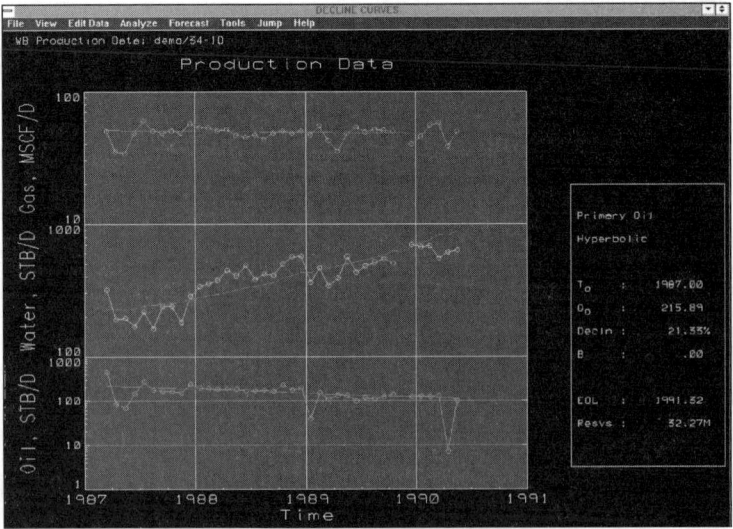

Fig. 12-5 • Oil, Water, and Gas Production

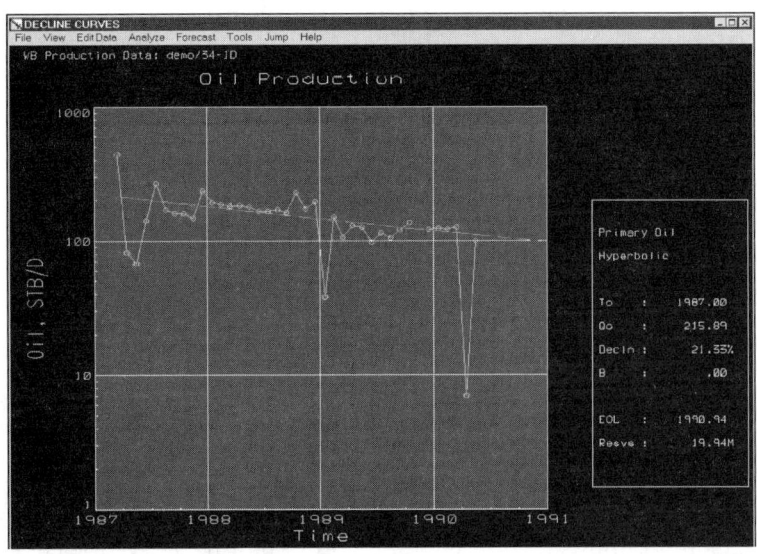

Fig. 12-6 • Oil Rate History Match and Prediction

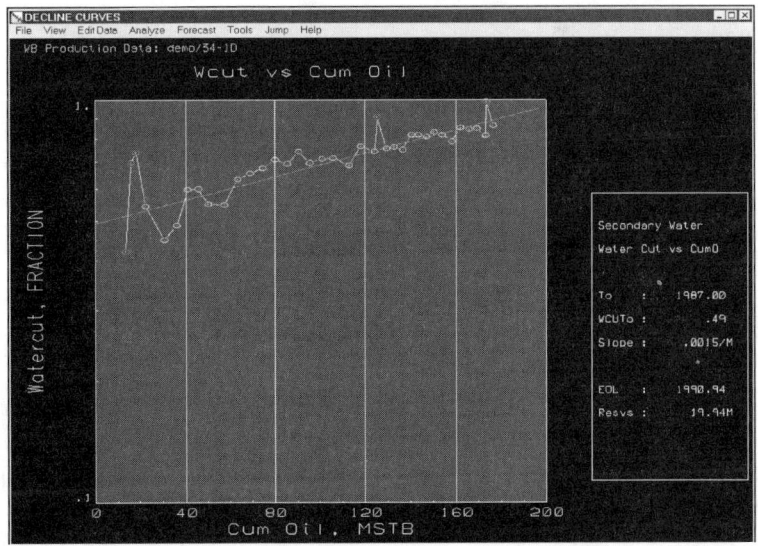

Fig. 12-7 • Water Cut History Match and Prediction

5 shows oil, water, and gas production performance of a well. Figure 12-6 shows performance prediction. The analysis gave exponential decline of 21.3% per year. Based on an economic production rate limit of 10 STBO per day, the future reserves were calculated to be 19.9 thousand barrels of oil at year 1990.9. Figure 12-7 is a plot of water cut vs. cumulative oil production, which also includes the performance prediction. It shows the well is expected to exceed 90% water cut before the end of the prediction curve (which goes to the economic production rate limit). It must be emphasized that both oil rate and water cut should be analyzed to ensure reliability in the results. In this example, a water cut limit of 90% would shut the well in, and the economic rate limit would not be a factor.

Ershaghi and Omoregie method

Because extrapolation of the past water cut plot is often complicated, Ershaghi and Omoregie[3] devised a method to plot recovery efficiency vs. X, as defined below, which yielded a straight line:

$$E_R = mX + n \qquad (12\text{-}13)$$

where:

E_R = over-all recovery efficiency

$$X = \ln(1/f_w - 1) - 1/f_w \qquad (12\text{-}14)$$

f_w = fraction of water flowing
m = slope
n = constant

This method is claimed to be more general than the classical plot of water cut vs. cumulative oil production, and more applicable when water cut exceeds 0.5. Given actual water cut vs. recovery efficiency data, a graph of recovery vs. X would result in a straight line, which may be extrapolated to any desired water cut to obtain corresponding recovery. The parameters m and n in Equation (12-13) can be derived from the straight-line relationship in Figure 12-8. These values then can be used in Equation (12-13) to predict

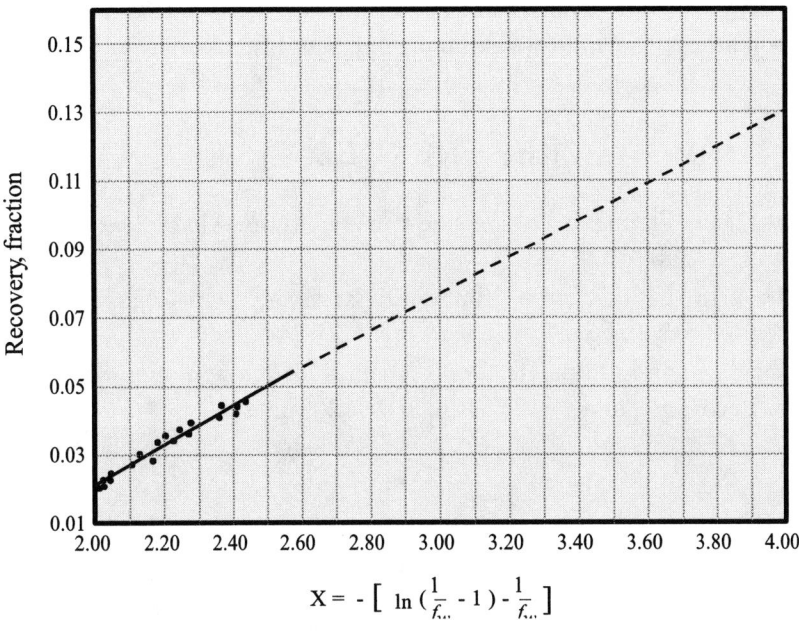

$$X = - \left[\ln \left(\frac{1}{f_w} - 1 \right) - \frac{1}{f_w} \right]$$

Fig. 12-8 • Recovery vs. X

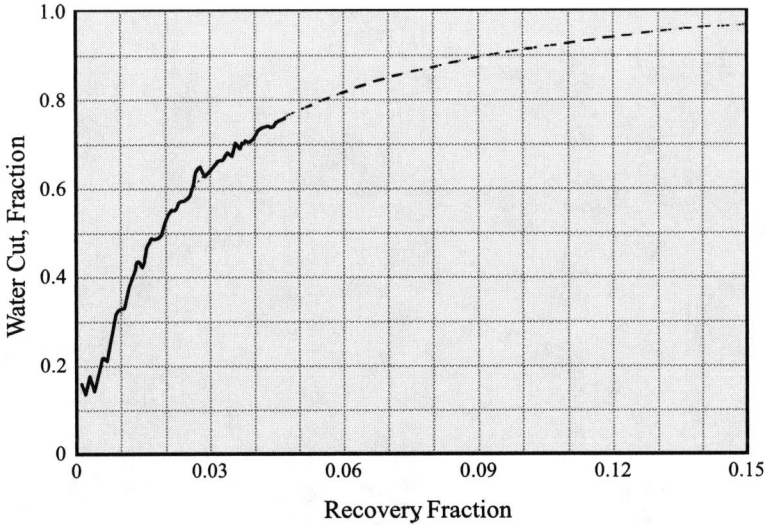

Fig. 12-9 • Recovery vs. Water Cut

water cut vs. oil recovery in Figure 12-9, which shows clearly that straight-line extrapolation of water cut would result in pessimistic ultimate recovery.

REFERENCES

1. Arps, J. J.: "Estimation of Decline Curves," Trans. AIME 160 (1945): 228-247

2. Arps, J. J.: "Estimation of Primary Oil Reserves," Trans. AIME 207 (1956): 182-194

3. Ershaghi, I. and Omoregie, O. "A Method for Extrapolation of Cut vs. Recovery Curves", JPT (Feb. 1978) 203-204

Chapter THIRTEEN

MATERIAL BALANCE METHOD

The classical material balance method is more fundamental than the decline curve technique for analyzing reservoir production performance. This method is used to estimate the original hydrocarbon in place (oil or gas), and ultimate primary recovery from a reservoir. It is based upon the law of conservation of mass, which simply means that fluids are conserved, *i.e.*, neither created nor destroyed. The basic assumptions made in this technique are:

- Homogeneous tank model, *i.e.*, rock and fluid properties are the same throughout the reservoir
- Fluid production and injection occur at single production and single injection points.
- There is no direction to fluid flow

However, in reality the reservoirs are not homogeneous, production and injection wells are areally distributed and activated at different times, and fluid flows in definite directions. Nevertheless, the material balance method is widely used and it is found to be a very valuable tool for reservoir analysis, with reasonably acceptable results.

The material balance equations for oil and gas reservoirs are used for the following:

- history match of past performance
- estimating original hydrocarbons in place
- prediction of future performance

Satter and Thakur[1] presented general material balance equations for oil and gas reservoirs, and discussed their applications[2,3,4,5,6,7].

OIL RESERVOIRS

General material balance for oil reservoirs can be expressed as the equation of a straight line:

$$F = N(E_o + mE_g + E_{fw}) + W_e \qquad (13\text{-}1)$$

where:
F = production of oil, water, and gas, rb
N = original oil-in-place, stb
E_o = expansion of oil and original gas in solution, rb/stb
m = initial gas cap volume, fraction of initial oil volume
E_g = gas cap expansion, rb/stb
E_{fw} = connate water expansion and pore volume reduction due to production, rb/stb
W_e = cumulative natural water influx, rb

The above equation contains three unknowns: original oil in place, gas cap size, and natural water influx. The unknown parameters can be determined by history matching for different drive mechanisms:

- *Solution gas drive* oil reservoirs—unknown is the original oil in place
- *Gas cap drive* oil reservoirs—unknowns are original oil in place and gas cap size

- *Water drive* oil reservoirs—unknowns are original oil-in-place and natural water influx

There are five commonly used graphical methods[2,3,4,5,6] based on the material balance straight-line equation:

1. FE method—plot of total production vs. total expansion
2. Gas cap method—plot of total production/oil expansion vs. gas expansion/oil expansion
3. Havlena-Odeh method—total production/total expansion vs. water influx/total expansion
4. Campbell method—total production/total expansion vs. total production
5. Pressure method—pressure vs. oil production

A more detailed discussion of the above methods may be found in reference 1.

Reservoir data required for history match are:

1. cumulative gas, oil, and water production from the reservoir for a series of time "points"
2. average reservoir pressures at the corresponding time "points" of (1)
3. PVT data for the reservoir fluids over the expected reservoir pressure ranges

If the original oil in place, gas cap size and aquifer strength and size are known, material balance equations can be used to predict the future performance. However, solutions for gas cap drive and natural water drive reservoirs are very complex. Performance prediction above the bubble point is rather straightforward. However, prediction below the bubble point requires simultaneous solutions of material balance equation and subsidiary liquid saturation, produced gas-oil ratio, and cumulative gas production.

GAS RESERVOIRS

General material balance as an equation of straight line for gas reservoirs can be expressed as:

$$F = G (E_g + E_{fw}) + W_e \qquad\qquad (13\text{-}2)$$

where:

G = original gas-in-place, scf

F = Production of gas, and water, rb

E_g = Expansion of gas, rb/scf

E_{fw} = Connate water expansion and pore volume reduction, rb/scf

W_e = cumulative natural water influx, rb

The unknown original gas in place, and natural water influx parameters in the above equation can be determined by history matching for depletion drive and water drive reservoirs. There are four commonly used graphical methods to calculate original gas in place:[2,3,4,7]

1. p/z method—reservoir pressure/z-factor vs. cumulative gas production
2. Cole method—total production/total expansion vs. total production
3. Havlena and Odeh method—total production/total gas expansion vs. water influx/total expansion
4. Pressure method—reservoir pressure vs. cumulative gas production

Data required for history match are:

1. cumulative gas, condensate, and water productions from the reservoir for a series of time "points"
2. average reservoir pressures at the corresponding time "points" of (1)
3. PVT data for the reservoir fluids over the expected reservoir pressure ranges

The future performance of gas reservoirs without water influx and water production can be calculated directly from a material balance equation giving a straight-line relationship between p/z and cumulative gas production. Data needed for calculating gas production are: gas compressibility (z) vs. pressure (p) and original gas in place. Performance prediction for a water drive gas reservoir is more complex.

MATERIAL BALANCE EXAMPLE—CASE STUDY

Satter and Thakur[1] reported computer analyses of several oil and gas reservoir case studies. Readers are referred to that book for details of these studies. The oil reservoir cases presented were an undersaturated reservoir, a gas cap drive reservoir, and a natural water drive reservoir with a small gas cap. The gas reservoir cases presented were a depletion drive Gulf Coast reservoir, an abnormally pressured reservoir, and a reservoir with an infinite linear aquifer.

An example of a material balance study carried out on a Middle East oil reservoir is presented in this chapter. The reservoir is Devonian in age and has the following general characteristics:

- Reservoir was discovered and placed on production in the early 1960s
- Approximately 30 years of production and pressure history were available
- Reservoir depth = approximately 7,700 ft.
- Initial temperature = 200° F
- Reservoir was saturated, with an initial pressure and bubble point of approximately 3,200 psia
- Oil gravity = approximately 42° API
- GOR = approximately 1,100 STB/SCF
- Connate water saturation = 12%
- Residual oil saturation estimated at 25%

Figure 13-1 shows historical pressure data available over a 30-year period. Average reservoir pressure was clearly declining for almost 20 years when an apparent re-pressuring started to occur, possibly due to delayed water influx into the reservoir. There was no record of water injection in this reservoir or in any other reservoir where the completed wells had penetrated. There were also no signs of any communication behind pipe, which could have accounted for the average reservoir pressure starting to increase slowly. Figures 13-2 and 13-3 show performance plots for daily oil, water, and gas production rates and water-cut and GOR, respectively.

The initial reservoir description, based upon well log analysis, showed a reservoir with an apparently large gas cap and a thin oil rim, with a large aquifer present. A decision was made to carry out a material balance study as part of an overall evaluation. MBAL, a general material balance program, from Petroleum Experts was the software package that was used in this analysis. (For other general material balance software packages available on

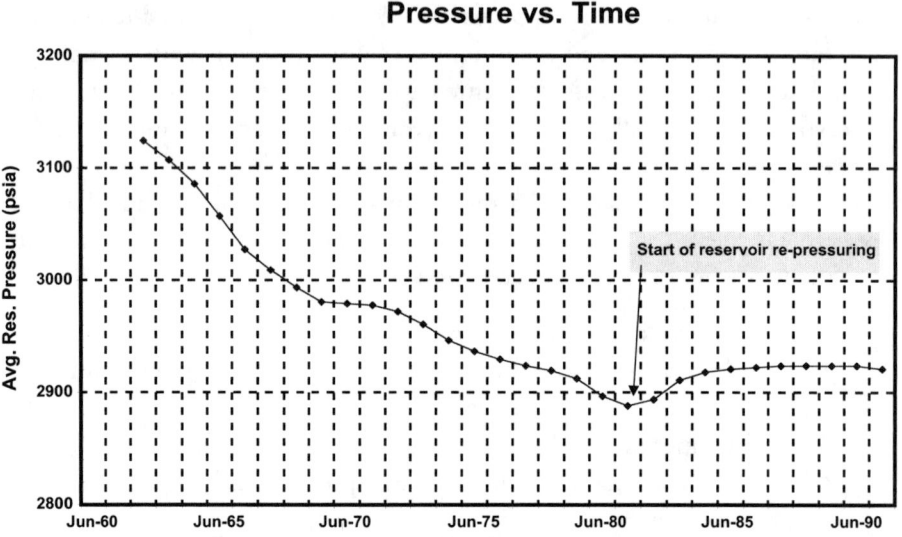

Fig. 13-1 • Pressure Data for Case Study

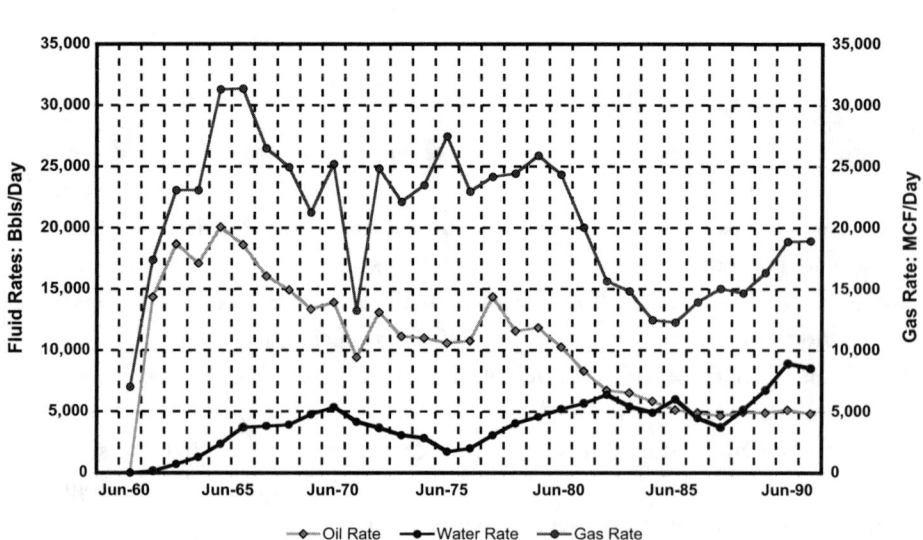

Fig. 13-2 • Production Rates for Case Study

the market, see chapter 15). The purpose of this material balance study was to obtain initial oil-in-place and the ratio of gas cap volume to oil volume. In addition, the relative strength of the aquifer would be determined.

Once all production data and pressure history were loaded into the MBAL program, regression calculations were carried out that varied parameters such as:

- aquifer model
- aquifer system (defines the flow prevailing in the reservoir and aquifer system, such as bottom water and edge water)
- aquifer boundary condition (constant pressure, infinite acting, or a sealed)
- reservoir equivalent radius
- aquifer permeability
- aquifer volume

The regression was carried out until an acceptable match was obtained

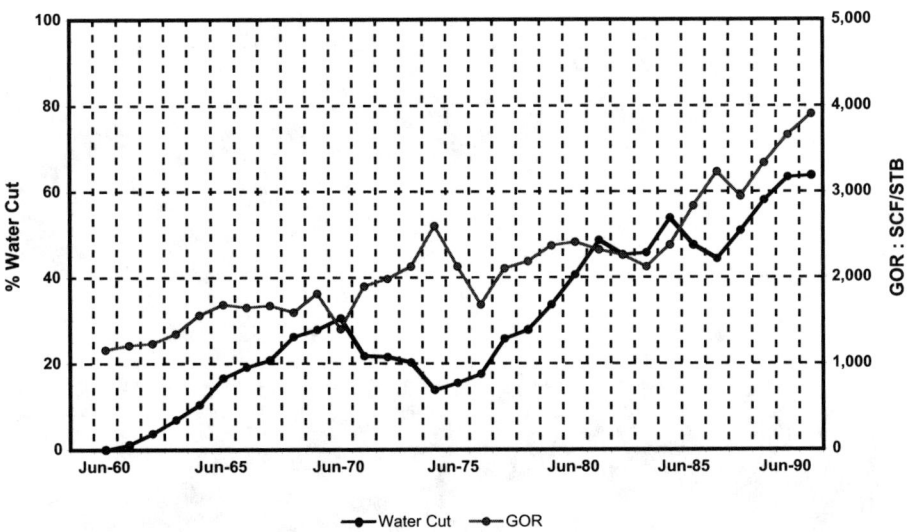

Water Cut & Gas-Oil-Ratio vs. Time

Fig. 13-3 • Water Cut and GOR for Case Study

to the pressure history. MBAL also allowed the regression "best fit" to be compared with other traditional material balance methods such as Havelena and Odeh and others. Figure 13-4 is a drive mechanism plot which shows how this reservoir started out with a dominant gas cap expansion. Approximately halfway through the 30+ year producing period, water influx became the dominant drive mechanism to the extent that at the end of history the average reservoir pressure had begun to slightly increase. There had been no reported gas or water injection for pressure maintenance. Figure 13-5 shows the result of regression history with pressure. The "best fit" regression gave the following results:

Aquifer model . . . . Hurst-Van Everdingen-Dake
Aquifer system . . . . Bottom water drive
Boundary model . . . . . . Constant pressure boundary
Aquifer volume . . . . 104,448 MM ft³
Aquifer permeability . . . . 375 md
Equivalent tank radius . . . . 1,500 ft.

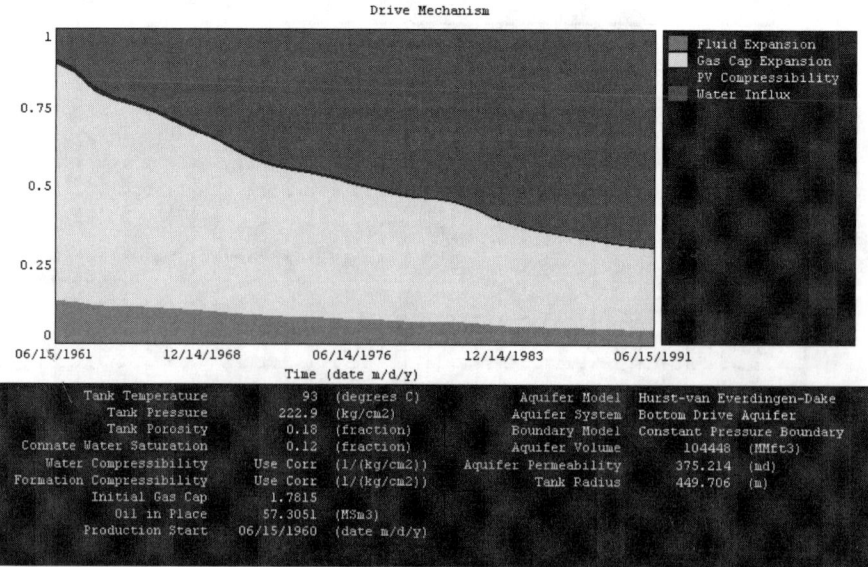

Fig. 13-4 • Drive Mechanism Plot

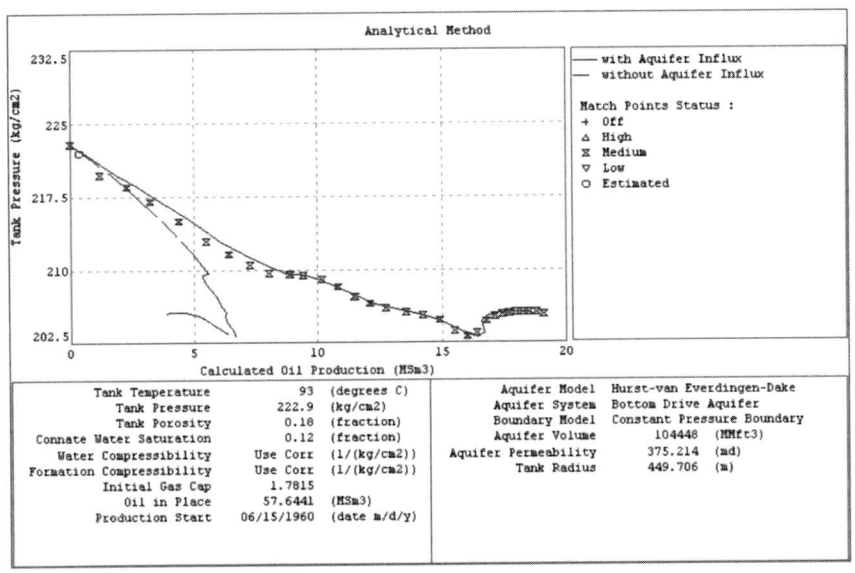

Fig. 13-5 • Regression on Pressure History

Initial oil in place . . . . 360 MMSTB oil
Gas cap/oil volume ratio . . . . 1.78 : 1

REFERENCES

1. Satter, A., and Thakur, G. C.: *Integrated Petroleum Reservoir Management: A Team Approach*, PennWell Books, Tulsa, Oklahoma (1994)
2. Havlena, D., and Odeh, A. S.: "The Material Balance as an Equation of Straight Line," J. Pet. Tech. (August 1963) 896-900
3. Havlena, D., and Odeh, A. S.: "The Material Balance as an Equation of Straight Line-Part II, Field Cases." J. Pet. Tech. (July 1964) 815-822
4. Wang, B., and Teasdale, T. S.: "GASWAT-PC: A Microcomputer Program for Gas Material Balance with Water Influx," SPE Paper 16484, Petroleum Industry Applications of Microcomputers, Del Lago on Lake Conroe, Texas, June 23-26, 1987

5. Campbell, R. A., and Campbell, J. M., Sr.: Mineral Property Economics, Vol. 3: Petroleum Property Evaluation, Campbell Petroleum Series (1978)

6. Schilthuis, R. J.: "Active Oil and Reservoir Energy," Trans. AIME (1936) 118, 33-52

7. Cole, F. W.: Reservoir Engineering Manual, Gulf Publishing Company (1969)

Chapter FOURTEEN

RESERVOIR SIMULATION

RESERVOIR SIMULATION

Reservoir simulators are widely used to study reservoir performance and to determine methods for enhancing the ultimate recovery of hydrocarbons from the reservoir. They play a very important role in the modern reservoir management process, and are used to develop a reservoir management plan and to monitor and evaluate reservoir performance during the life of the reservoir, which begins with exploration leading to discovery, followed by delineation, development, production, and finally abandonment. Satter and Thakur[1] provide an overview of reservoir simulation concepts, and processes.

Concept and well management

Numerical simulation is based upon the following concepts:

- Material balance principles
- Reservoir heterogeneity
- Direction of flow
- Flow of phases
- Spatial distribution of wells

Unlike the classical material-balance approach, a reservoir simulator incorporates details of well management:

- Location of production and injection wells
- Completions
- Rate or bottom hole pressure specified, or even both

The wells can be turned on or off at desired times with specified downhole completions. The well rates and/or the bottom hole pressures can be set as desired.

Technology and spatial and time discretization

Simulation has become a reality because of advances made in computing power and the technology now available for data handling, computational techniques, report writing, and graphics. These allow the reservoir to be divided into many small tanks, cells, or blocks to take into account heterogeneity (Fig. 14-1). Computations of pressures and saturations for each cell are carried out at discrete time steps, starting with the initial time.

Phase and direction of flow

Black oil simulators are characterized by the number of (fluid) phases, directions of flow, and the type of solution used for the complex fluid flow equations. The fluid phase can be:

- single phase (oil or gas)
- two phase (oil and gas, or oil and water)
- three phase (oil, gas, and water)

The direction of flow can be:

- 1-Dimensional linear or radial, when only in one direction

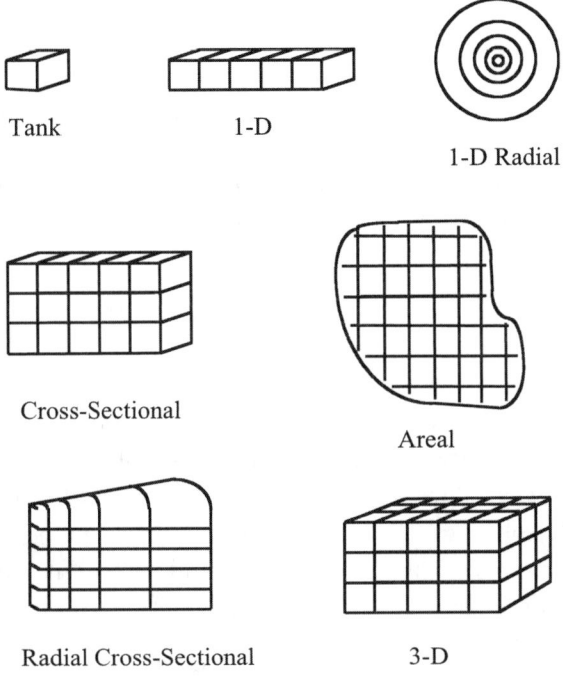

Fig. 14-1 • Typical Simulation Models

- 2-Dimensional areal, cross-sectional, or radial cross-sectional when in x-y, x-z, or r-z directions
- 3-Dimensional when in x-y-z direction

Fluid flow equations

Simultaneous flow of oil, gas, and water through a porous medium is described by a set of three (oil, gas, water) complex partial differential equations:

- Law of conservation of mass
- Fluid flow law (Darcy's)
- PVT behavior of fluids

These flow equations contain six unknowns:

$$S_o, S_g, S_w \text{ and } p_o, p_g, p_w$$

where S and p denote saturation and pressure, respectively, and subscripts o, g, and w denote oil, gas, and water.

Auxiliary relationships involving saturations and capillary pressures must be used in order to solve the fluid flow equations:

$$S_o + S_g + S_w = 1$$
$$p_{cow} = p_o - p_w = p_{cow}(S_o, S_w)$$
$$p_{cog} = p_g - p_o = p_{cog}(S_o, S_g)$$

where p_{cow} and p_{cog} denote capillary pressure between oil-water and oil-gas, respectively. If the capillary pressure is zero, there is only one pressure, so the number of unknowns reduces to pressure, and oil, gas, and water saturations.

Approximate solutions of the complex equations can be obtained by using finite difference schemes. The pressures and saturations can be solved explicitly, implicitly, or by a combination method:

- Explicit—explicit pressure, explicit saturation
- Implicit—implicit pressure, implicit saturation
- Combination—implicit pressure, explicit saturation (IMPES)

RESERVOIR SIMULATORS

Reservoir simulators are generally classified in four categories governed by flow mechanisms:

Black Oil: Fluid flow

Compositional: Fluid flow
                  Phase composition

Thermal: Fluid flow
                  Heat flow

Chemical: Fluid flow
Mass transport due to dispersion, adsorption,
and partitioning

Various stand-alone and integrated software packages for black oil, compositional, and thermal models are reported in the next chapter.

SIMULATION PROCESS

In general, the reservoir simulation process (Fig. 14-2) can be divided into three main phases:

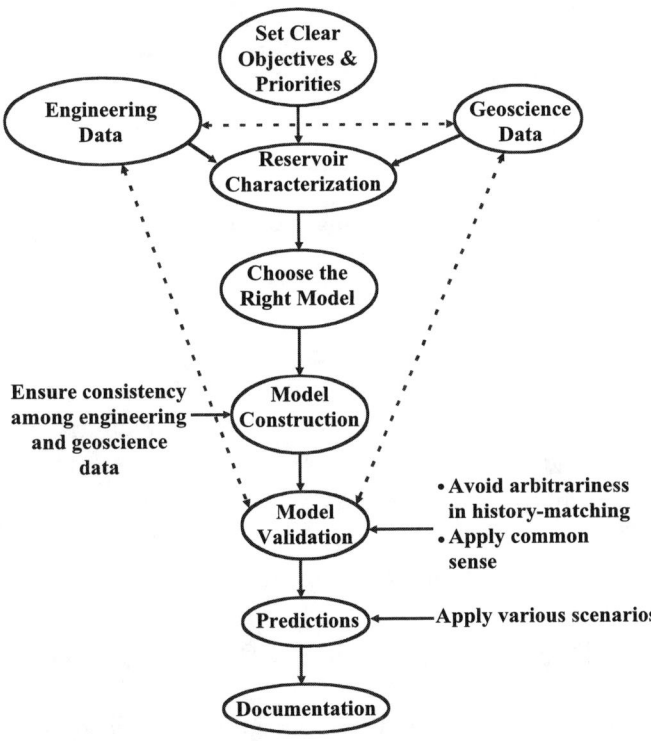

Fig. 14-2 • Reservoir Simulation (After SPE paper 18305 by Saleri and Toronyl)

- *Input Data Gathering* - geological, reservoir, well completions, production, injection, etc.
- *History Matching* - initialization, pressure match, saturation match, and productivity index match
- *Performance Prediction* - existing operating and/or some alternative development plan

Input data consists of general data, grid data, rock and fluid data, production/injection data, and well data. Gathering the needed data can be very time consuming and expensive. Ascertaining the reliability of the available data is vital for successful reservoir modeling.

History matching of the past production and pressure performance consists of adjusting the reservoir parameters of a model until the simulated performance matches the observed or historical behavior. This is a necessary step before the prediction phase because the accuracy of a prediction can be no better than the accuracy of the history match. However, it must be recognized that history matches are not unique.

The stepwise history matching procedure[2] consists of:

- pressure matching (Fig. 14-3) followed by
- saturation matching (Fig. 14-4) followed by
- productivity matching (Fig. 14-5)

Predicting future performance of a reservoir under existing operating conditions and/or some alternate development plan such as infill drilling, waterflood after primary, etc., is the final phase of a reservoir simulation study. The main objective is to determine the optimum operating condition in order to maximize economic recovery of hydrocarbons from the reservoir.

ABUSE OF RESESRVOIR SIMULATION

The use of reservoir simulation has grown steadily over the last 25 years because of the constant improvement in simulator software and computer hardware. The rapid growth and acceptance of simulation has led to some confusion and occasional misuse of the reservoir engineering tool because of

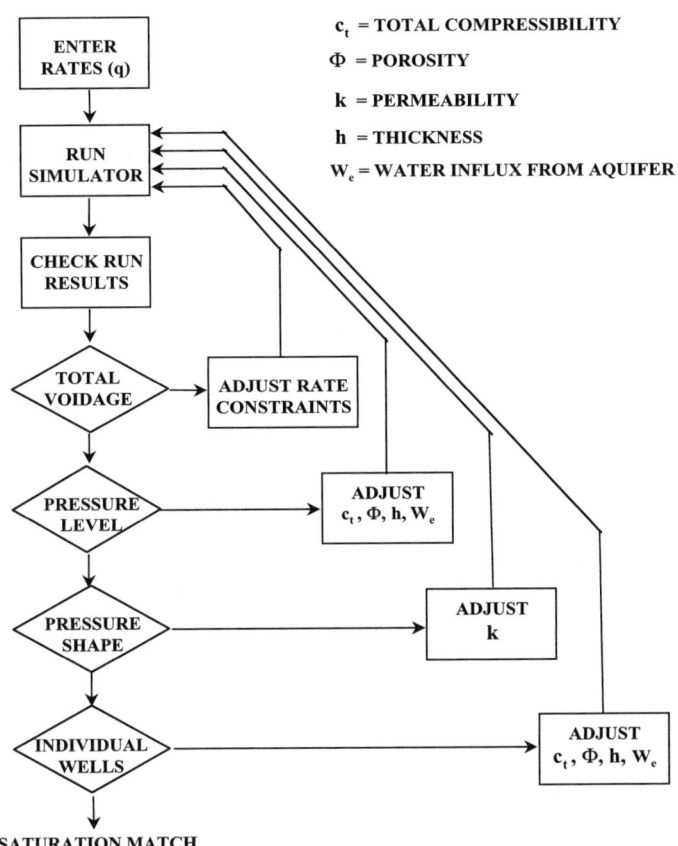

Fig. 14-3 • Pressure Match Procedure

unrealistic expectations, insufficient justification for simulation, and unrealistic reservoir description.

GOLDEN RULES FOR
SIMULATION ENGINEERS

Recognizing that there is increasing danger that inexperienced users will

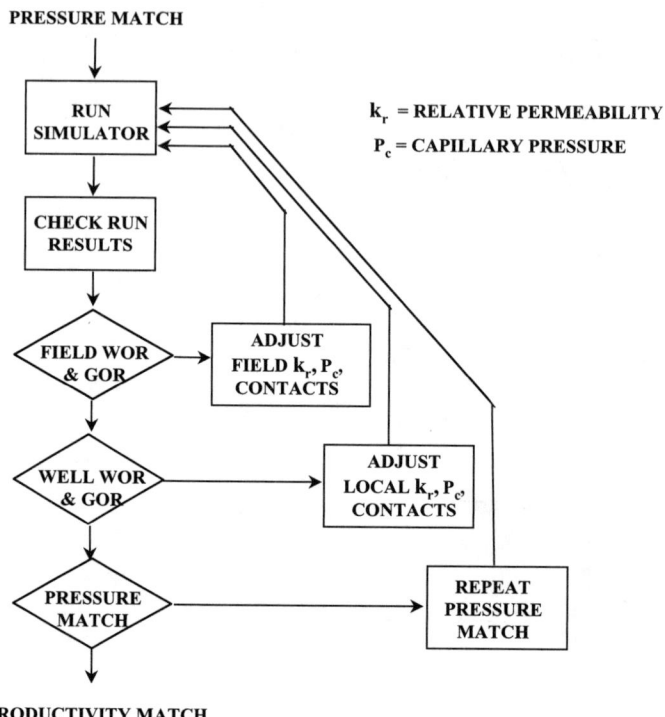

PRESSURE MATCH

RUN SIMULATOR

CHECK RUN RESULTS

FIELD WOR & GOR

ADJUST FIELD k_r, P_c, CONTACTS

WELL WOR & GOR

ADJUST LOCAL k_r, P_c, CONTACTS

PRESSURE MATCH

REPEAT PRESSURE MATCH

k_r = RELATIVE PERMEABILITY
P_c = CAPILLARY PRESSURE

PRODUCTIVITY MATCH

Fig. 14-4 • Saturation Match Procedure

misuse sophisticated models available to them, Aziz[3] offered the following basic rules in order to minimize this danger:

1. Understand your problem and define your objectives.
2. Keep it simple. Start and end with the simplest model. Understand model limitations and capabilities.
3. Understand the interaction between different parts of the model. Reservoir, aquifer, wells, and facilities are interrelated.
4. Don't assume bigger is always better. Always question the size of a study that is limited by the computer resources and/or the budget. Quality and quantity of data are important.

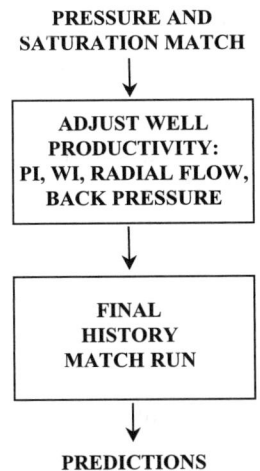

Fig. 14-5 • Productivity Match Procedure

5. Know your limitations and trust your judgment. Remember that simulation is not an exact science. Do simple material balance calculations to check simulation results.
6. Be reasonable in your expectations. Often the most you can get from a study is some guidance on the relative merits of choices available to you.
7. Question data adjustments for history matching. Remember that this process does not have a unique solution.
8. Never smooth or average out extremes.
9. Pay attention to the measurements and the scales at which they were made. Measured values at the core scale may not directly apply at the larger block scale, but measurements at one scale do influence values at other scales.
10. Don't skimp on necessary laboratory data. Plan laboratory work with its end use in mind.

EXAMPLES

Applications of Baker-Hughes/SSI's Black Oil Simulator are presented for:

- Newly discovered field development plan in chapter 16
- North Apoi/Funiwa field optimization study in chapter 17
- Waterflood project development in chapter 18

REFERENCES

1. Satter, A. and Thakur, G. C.: *Integrated Petroleum Reservoir Management: A Team Approach*, PennWell Books, Tulsa, Oklahoma (1994)
2. *Petroleum WorkBench Reference Manuals*, Scientific Software Intercomp, Denver, Colorado
3. Aziz, Khalid: "Ten Golden Rules for Simulation Engineers," Dialog, JPT, Nov., 1989

Chapter *FIFTEEN*

COMPUTER SOFTWARE

In the early days of computerization, most of the software available to the geoscientist or engineer consisted of mainframe applications that ran in batch mode. The nature of these programs generally restricted their use to a few specialists at the larger companies. Over the years, the evolution of both hardware and software technology has provided many new tools for improving the reservoir management process. Speed of operation, data capacity, and communication capability, which have improved exponentially while costs have decreased, have brought personal computers or workstations to nearly every desk. Operating systems and the applications themselves have become easier to use, with emphasis on reaching a wide variety of users, not just the specialist.

Many of the major oil companies wrote their own software in the past, seeking a competitive advantage for their proprietary techniques. Although this is still done to some extent, companies now have the need of so many applications that it is not cost effective to create them all in-house and to provide support to the growing number of users. This has led to the growth of a large service industry, which develops and markets software throughout the oil and gas business.

STAND-ALONE SOFTWARE

Many computer applications have been developed for specific purposes. It is faster and more economic to produce a program to focus on a specific task. Such programs are often easy to use and relatively inexpensive. The simplest of these stand-alone programs may still be written by the end-user, or at least within the user's company (such as a spreadsheet to do a specific type of log analysis). More complex applications come from the service companies (which may also offer consulting services based on their own use of the software) and software specialty companies. Table 15-1 lists a few of the many stand-alone computer programs currently on the market, which play an important role in our reservoir management process. Nearly all of the listed software programs are either Windows- or Unix-based applications. This list is by no means intended to be all-inclusive, but it is representative of the types of stand-alone software packages that are available today.

INTEGRATED SOFTWARE PACKAGES

As more and more reservoir management tasks have become computerized, the need has grown to use the results of one program as input to another. Several vendors have developed "integrated packages" by combining several applications in such a way that data can be easily moved from one program to another. All programs within one package generally have a similar look and feel, or common user interface, making it easier for one user to operate all of them. To date, few such packages (if any) have been developed as a cohesive unit. Instead, they have been created by patching together several previously existing stand-alone applications. There are several service

Vendor	Program	Description
D. P. COOK	OGRE	Economic evaluation program
EPS	PANSYSTEMS	Well test design and pressure transient analysis
	ESP	Submersible pump design and analysis
	WELLFLO	Nodal analysis and gas lift design and optimization
FEKETE	PIPER	Gas pipeline, wellbore, and reservoir deliverability model
	WELLTEST	Pressure transient analysis
	VALIDATA	Read pressure and temperature data from electronic recorder
	FIELDNOTES	Transfer well production test data from field to office
INTERA	SIMTRAN	Numerical well test analysis
	SIMFRAC	Design of propped and acid fracturing treatments
	TERASIM	Advanced reservoir simulation software
	SIMPERF	Near wellbore simulator to optimize perforation performance
	GEOSTAT	Transfer fine-scale geostatistical results into reservoir simulators
KAPPA	SAPHIR	Pressure transient analysis
	EMERAUDE	Production logging analysis
MERAK	PEEP	Economic evaluation program
	WELLVIEW	Well completion data management
	FORECAST	Production decline and P/Z analysis
	DEC. TREE	Petroleum decision and risk analysis
	FDM	Field production management
PETROLEUM EXPERTS	MBAL	General material balance
	PROSPER	Nodal analysis
	GAP	Production networks and gas lift optimization
PI DWIGHT'S	OILWAT/GAS WAT	General material balance for oil and gas reservoirs
	VOLRES	Volumetric determination of oil and gas reserves
	SUBPUMP	ESP design and analysis
	PERFORM	Wellbore nodal analysis
	PIPESOFT II	Gas surface network gathering system
RDXAR	MORE	Black oil and compositional reservoir simulator
	IRAP-RMS	Geological description
T. TAN	EXODUS	PC-based reservoir simulator

Table 15-1 • Stand-alone Software Packages

companies that have a first-generation set of integrated reservoir management software. Table 15-2 summarizes these various applications. Again, this is not intended to be a complete list.

Most service companies originated around a fairly narrow range of expertise and have not had the capability to create truly all-encompassing integrated packages. This has resulted in numerous mergers of companies whose strengths complimented each other.

Vendor	Program	Description
BAKER	WORKBENCH	Contains a number of modules
ATLAS	BLACK OIL	Black oil reservoir simulation
GEOSCIENCE	RESERVOIR DESCRIPTION	Reservoir characterization, logging, and geostatistics
(SSI)	PDA	Decline curve analysis
	INTERPRET	Well test design and analysis
	WPM	Well nodal analysis
	THERM	Thermal reservoir simulation
	COMP V	Compositional reservoir simulation
	PVT	Equation of state fluid PVT property
COMPUTER	IMEX	Reservoir simulation
MODELLING	STARS	Thermal simulation
GROUP	WINPROP	Equation of state PVT fluid property determination
	GEM	Compositional reservoir simulation
	BUILDER	Simulation pre-processor (no log analysis)
	RESULTS	Simulation post-processing
GEOQUEST	FINDER	Data Management System
	IESX, CHARISMA	2D/3D/4D seismic interpretation
	GEOVIZ, GEOCUBE	3D visualization and interpretation
	GEOFRAME	POSC-compliant integration platform
LANDMARK	VIP	Suite of integrated packages
GRAPHICS	VIP-ENCORE	Black oil module
	VIP-THERM	Models hot water and steam injection
	DESKTOP-PVT	Interactive fluid phase behavior package
	GEOLINK	Geological-to-engineering model interface
	GRIDGENR	Interactive simulation grid generator
	2D/3D VIEW	Simulation visualization
	STRATAMODEL	Advanced reservoir characterization

Table 15-2 • Integrated Software Packages

PETROTECHNICAL OPEN
SOFTWARE CORPORATION

Data management has become one of the greatest difficulties in using this vast array of software tools available today. It is usually convenient to move data among the various programs within one integrated package or even among the stand-alone programs from one supplier. There has been limited standardization throughout the industry. This has made it very difficult for users to retrieve data from their company database (if they have one at all) and input it to various applications from different vendors.

Several years ago, a group consisting of most of the oil and gas companies and service companies got together to discuss this problem, which they all shared. The result was the formation of Petrotechnical Open Software Corporation (POSC), for the purpose of defining data storage and software standards for the entire industry.

Some progress has been made by POSC, but the legacy of existing software and databases within the major oil companies and service companies alike has hampered agreement and development of true standards. Hopefully, time will solve this problem, as older software becomes obsolete and needs to be replaced.

In the meantime, geoscientists and engineers can take advantage of the tools now available to gain a more thorough understanding of their reservoirs and to develop optimum plans for managing those assets.

COMPUTING FACILITIES

Computing facilities in the petroleum industry have changed greatly over the past few years. As hardware has become more powerful and less expensive, central mainframe computers have been replaced by personal computers and networked workstations. One modern example is the Reservoir Computing Facility at Texaco Exploration & Production Technology Department (EPTD), illustrated in Figure 15-1. These computers are often used by EPTD staff to run projects for Texaco's business units. They can also be accessed from the remote division offices over the network.

There are many software applications installed at EPTD in order to provide compatibility with the division offices and outside project partners. The principle applications are shown in Table 15-3.

INTERNET

Use of the WWW (World Wide Web) has exploded in the petroleum industry, with the service companies as well as many of the major oil companies setting up their own public web sites. For the vendors and service companies, these web sites supply a vast wealth of information such as:

• company mission statements

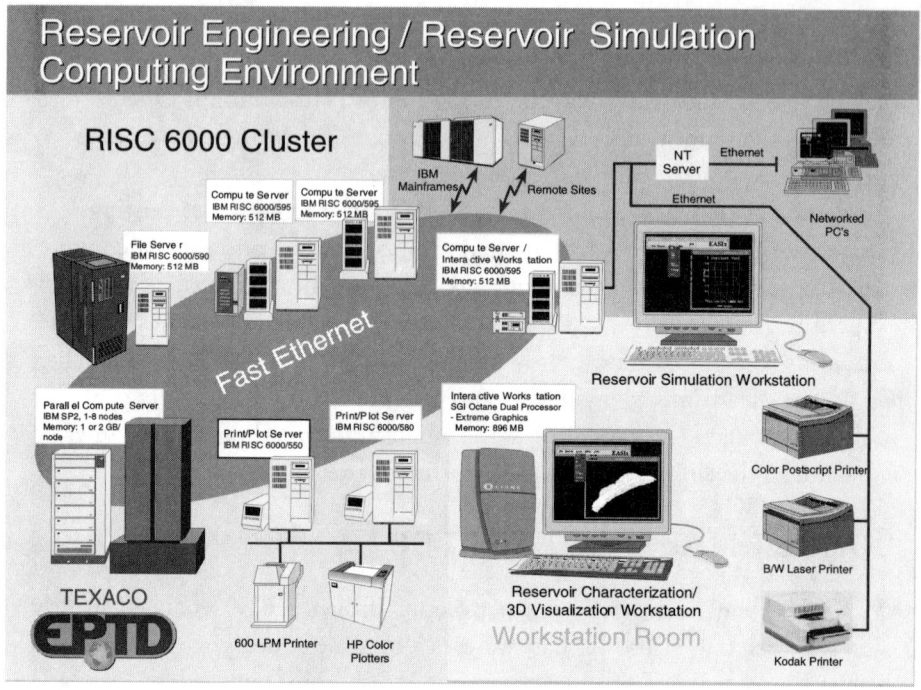

Fig. 15-1 • Texaco EPTD Computing Environment

- sales and marketing information
- demonstration versions of software programs
- service packs to patch and/or upgrade current versions of the software
- technical bulletins and product information
- phone numbers and e-mail addresses for support and sales contacts

Table 15-4 lists the URL addresses of several companies that have posted information about their products and organization on the Internet. A more complete listing of energy related links may be found at the web site of the Petroleum Technology Transfer Center—*www.pttc.org/linkindx.htm*. Many companies are increasingly using private, registration-only extranets, often on trusted servers, to provide secure, real-time project management service and communication with their customers.

Internet search engines, such as *www.yahoo.com*, can be used to expand

Model Type	Landmark	Baker Hughes/ SSI	Computer Modeling Group	University of Texas	GeoQuest Reservoir Technologies	Smedvig	Gemini	Texaco	Chevron
Black Oil	VIP-ENCORE	SIMBEST II	IMEX		ECLIPSE 100	MORE	MERLIN	UNISIM	CHEERS
Compositional	VIP-COMP	COMP III	GEM	UTCOMP	ECLIPSE 300	MORE			CHEERS
Thermal	VIP-THERM	THERM	STARS	UTTHERM					CHEERS
Chemical				UTCHEM					
Miscible	VIP-ENCORE	SIMBEST II	IMEX						
PVT	DESKTOP-PVT	PVT	CMGPROP		PVT				
Pre-Post Processors	PR-EXEC GRIDGENR HVWELL LGR PLOTVIEW 3DVIEW	AHM	XYPLOT RESULTS		VFPI GRID EDIT LGR/GC FILL GRAF RTVIEW		APPRENTICE		CHIMAP FLUVOL CHITUBE CHILOG
Integrated Applications	DESKTOP-PVT GEOLINK	WORKBENCH							
Reservoir Characterization	SIGMAVIEW SGM							GRIDSTAT	GEOLITH

Table 15-3 • Reservoir Computing Facility Installed Applications

Organization	Web Site URL
American Assoc. of Petroleum Geologists	www.aapg.org
American Petroleum Institute	www.api.org
Baker Hughes	www.bakerhughes.com
BJ Services	www.bjservices.com
Bureau of Economic Geology	www.utexas.edu/research/beg
Computer Modeling Group	www.cmgroup.com
Department of the Interior	www.mms.gov
Dresser	www.dresser.com
Exodus Software	www.petrostudies.com
Federal Energy Technology Center	www.fetc.doe.gov
Fekete Associates	www.fekete.com
Gas Research Institute	www.gri.org
Geoquest	www.slb.com/oilf/geoquest
Halliburton	www.halliburton.com
Kappa Engineering	www.kappaengineering.com
Landmark Graphics	www.lgc.com
Merak	www.merak.com
PennWell Publishing	www.pennwell.com
Petroleum Technology Transfer Council	www.pttc.org
PGS	www.pgs.com
PI Dwight's	www.ihsenergy.com
PTTC list of Links	www.pttc.org/linkindx.htm
Schlumberger	www.connect.slb.com
Scientific Software-Intercomp	www.ssii.com
ROXAR	www.smedtech.com
Society of Prof. Well Log Analysts	www.spwla.org
SPE Gulf Coast Section	www.spegcs.org
SPE International	www.spe.org
Sperry Sun	www.sperry-sun.com
U.S.G.S. home page	www.info.er.usgs.gov
University of Texas	www.pe.utexas.edu

Note: URLs sometimes change

Table 15-4 • Energy-Related Web Sites

this initial list and to pinpoint specific needs. The following are brief excerpts of information taken from the Internet about various companies and groups. They represent just a very small sample of the information available from the Internet. These excerpts are illustrated by figures copied from the associated home pages:

Baker Hughes (SSI) - Figure 15-2

Baker Atlas Geoscience subsidiary Scientific Software-Intercomp develops and markets software for development and production, for pipeline and surface facilities, and provides associated interdisciplinary technical support services, consulting, and training. One of its principal software products is Petroleum WorkBench.

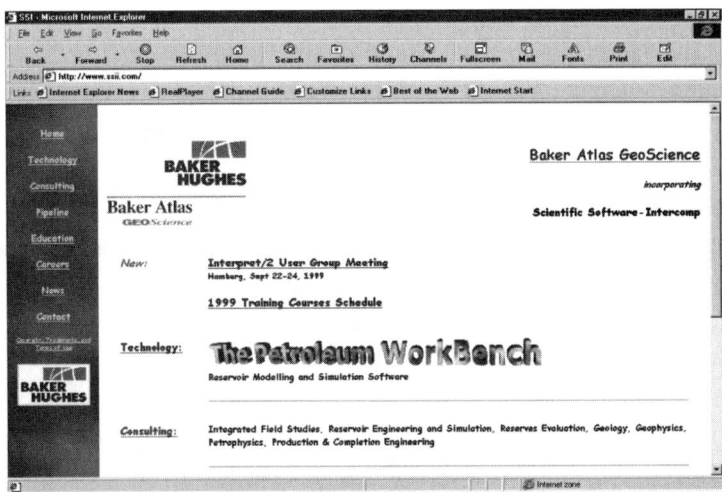

Fig. 15-2 • Baker Hughes (SSI)

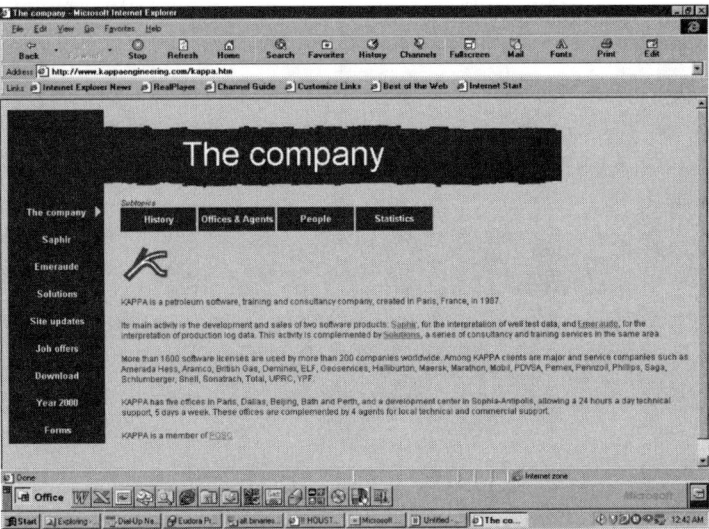

Fig. 15-3 • KAPPA Engineering

KAPPA Engineering - Figure 15-3

KAPPA Engineering is a software and consulting company. Primary software products are SAPHIR for pressure transient analysis and EMER-AUDE for production logging.

(RC)^2

Reservoir Characterization Research and Consulting, (RC)^2, deals with the integration of diverse data types for petroleum reservoir modeling and provides a wide-range of software tools for reservoir description as well as geostatistical modeling.

Fekete Associates - Figure 15-4

Fekete Associates is a Canadian petroleum engineering consulting firm that has software tools for well pressure transient analysis, economics, gas gathering systems, and production engineering.

Smedvig Technologies

Smedvig Technologies is a service company that provides multi-discipline software in the areas of basin modeling, petrophysics, gridding, reservoir simulation, and drilling, and has merged with Multi-Fluid.

CMG - Figure 15-5

Computer Modeling Group provides software for reservoir simulation of black-oil, compositional, and thermal reservoirs.

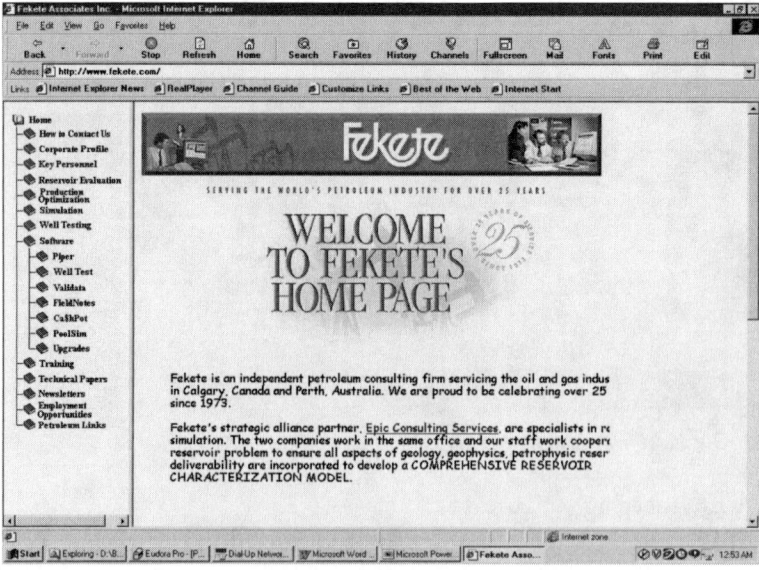

Fig. 15-4 • Fekete Associates

Beicip-Franlab - Figure 15-6

Beicip-Franlab is a French-based service company that offers software for basin analysis, reservoir characterization, reservoir simulation, and 3-D model building.

Fig. 15-5 • Computer Modeling Group (CMG)

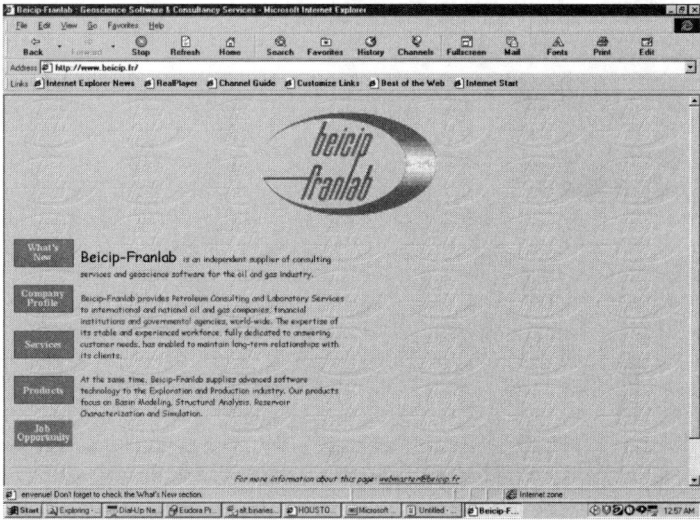

Fig. 15-6 • Beicip-Franlab

Landmark Graphics - Figure 15-7

Landmark Graphics, a Halliburton Company, provides integrated software for seismic interpretation, reservoir characterization, basin modeling, and reservoir simulation. Key products include SeisWorks, PetroWorks, Z-Map Plus, StrataModel, Desktop VIP, and 3D View.

U. S. Department of Energy - Figure 15-8

This is an example of reservoir management software developed in the public domain and available from the federal government.

Merak - Figure 15-9

Merak develops and markets Windows™ application software for the upstream energy industry. Merak's suite of integrated tools assists in the evaluation of exploration risk, production forecasts, economic projections, investment viability, and production operations. Features include mapping, budgeting, regime-specific financial projections, rig scheduling, wellbore schematics, reserves, and production data management.

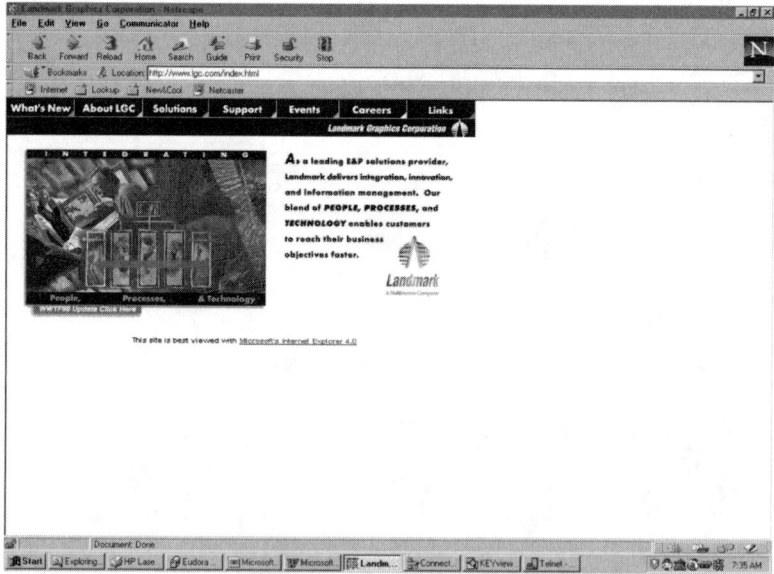

Fig. 15-7 • Landmark Graphics

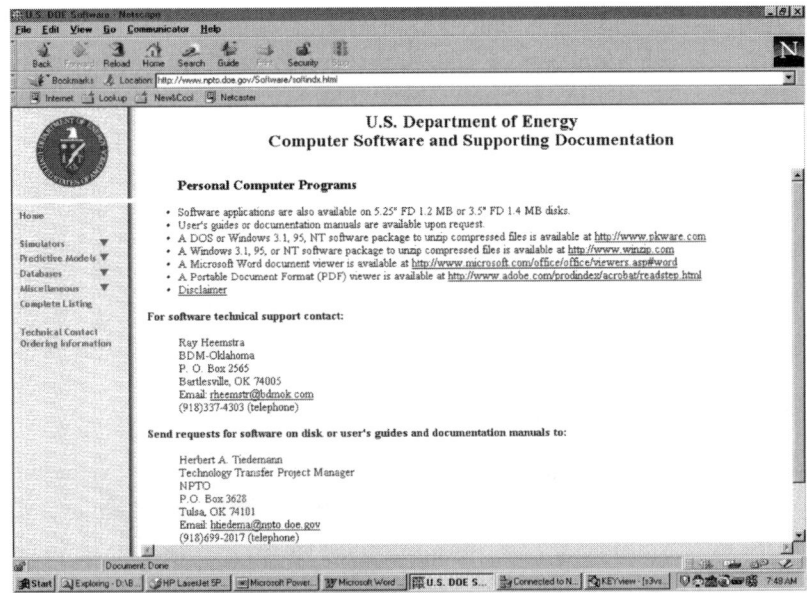

Fig. 15-8 • U.S. Department of Energy

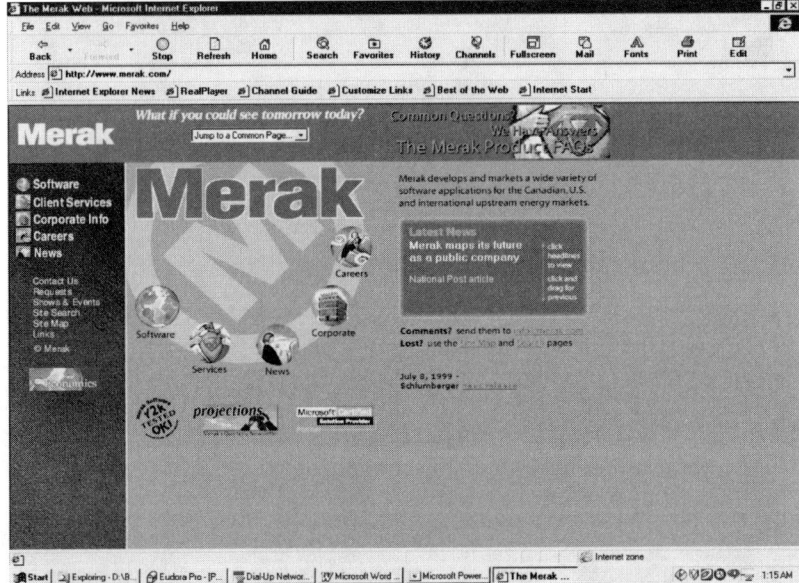

Fig. 15-9 • Merak

SPE Gulf Coast Section - Figure 15-10

This is the home page of the Society of Petroleum Engineers' Gulf Coast Section located in Houston, Texas. The web site offers a career management section, on-line event registration, and other services for its members and other energy professionals such as an in-depth group of energy related web links.

SPE International - Figure 15-11

This is the home page of the Society of Petroleum Engineers International. The web site offers on-line information, membership directory, industry directory with corporate links, technical paper search, plus links to local sections as well as an Internet Career Center.

Halliburton - Figure 15-12

Halliburton is a major oil field service company offering a full range of services to the energy industry. They are a world leader in energy equipment, energy services, engineering, and construction. Their goal is to provide their customers with solutions that increase oil and gas production while lowering costs.

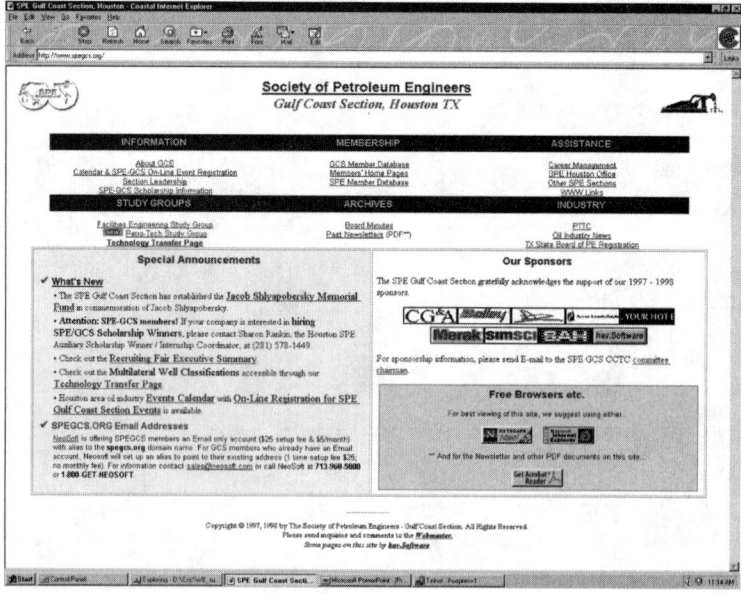

Fig. 15-10 • SPE Gulf Coast Section

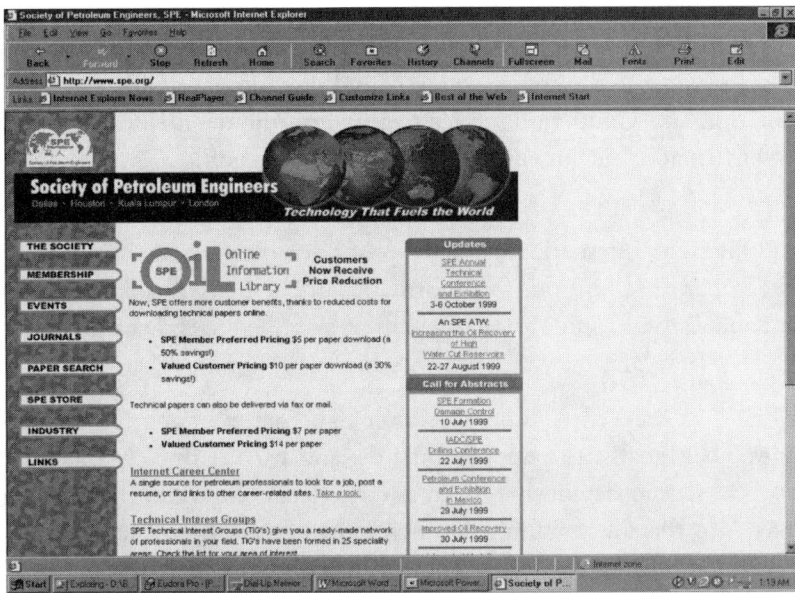

Fig. 15-11 • SPE International

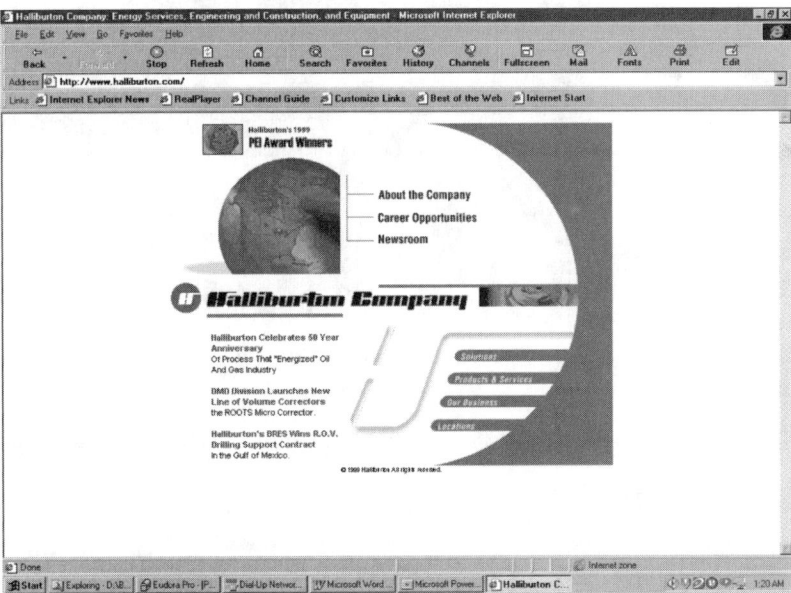

Fig. 15-12 • Halliburton

Schlumberger - Figure 15-13

Schlumberger is another major oil-field service company offering a full range of services to the energy industry and providing virtually every type of exploration and production service required during the life of an oil and gas reservoir (material included courtesy of Schlumberger).

POSC - Figure 15-14

POSC is an international not-for-profit organization whose mission is to provide open specifications for information modeling, information management, and data and application integration over the life cycle of E & P assets.

FETC - Figure 15-15

FETC is the Federal Energy Technology Center of the Department of Energy. Their goal is to perform, procure, and partner in technical research, development, and demonstration to advance technology into the commercial marketplace, thereby benefiting the environment, contributing to U.S. employment, and advancing the position of U.S. industries in the global market.

Fig. 15-13 • Schlumberger

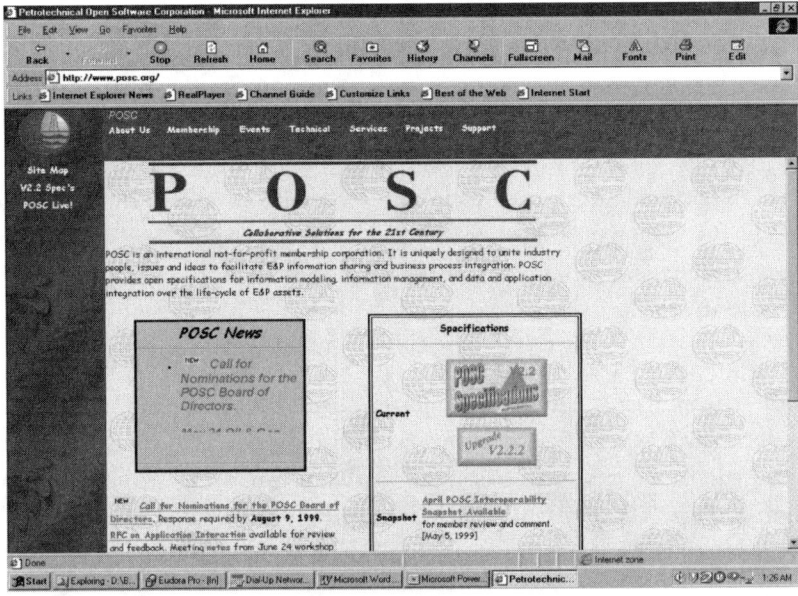

Fig. 15-14 • Petrotechnical Open Software Corporation (POSC)

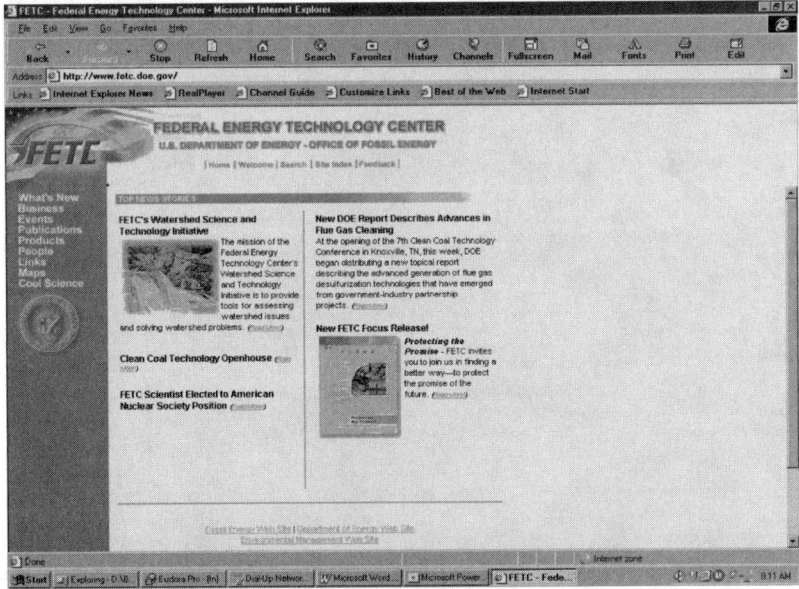

Fig. 15-15 • Federal Energy Technology Center (FETC)

Fossil Energy

This is the Office of Fossil Energy from the U.S. Department of Energy. This web site is becoming a primary means for communicating the progress and performance of the U.S. Department of Energy's fossil energy program. They manage a national technology program to increase natural gas and petroleum supplies and provide cleaner, more efficient ways to use coal and natural gas to generate electricity. They also oversee the Strategic Petroleum Reserve (the nation's emergency oil stockpile) and the Naval Petroleum and Oil Shale Reserves.

Edinburgh Petroleum Services - Figure 15-16

EPS' founding mission statement encapsulates the principle of providing the upstream oil and gas industry with cost effective solutions to harnessing the industry's resources. EPS' software products are based upon a commitment to research and development. EPS software is used to leverage skills through increasing productivity and by providing cutting edge technology that can be relied on.

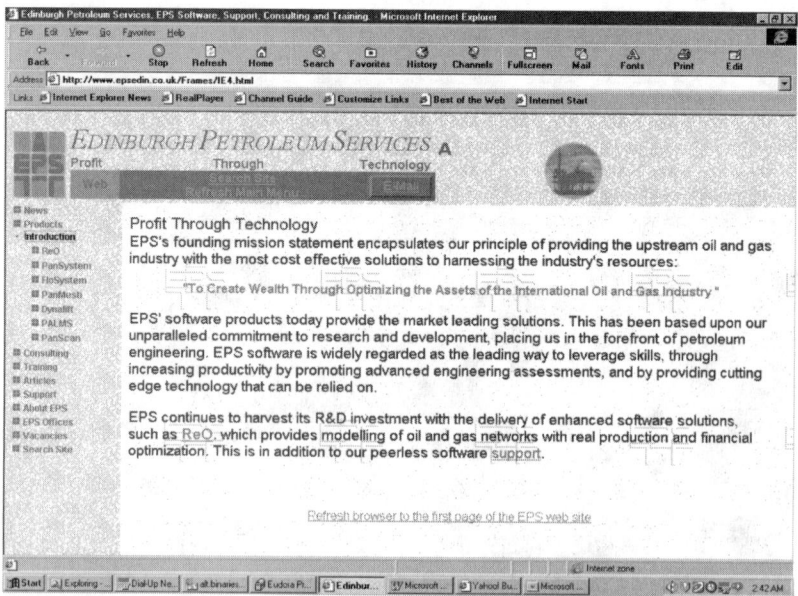

Fig. 15-16 • Edinburgh Petroleum Services (EPS)

Petroleum Technology Transfer Council - Figure 15-17

The Petroleum Technology Transfer Council (PTTC) was formed in 1994 by the U.S. oil and natural gas exploration and production (E&P) industry to identify and transfer upstream technologies to domestic producers. PTTC's technology programs help producers reduce costs, improve operating efficiency, increase ultimate recovery, enhance environmental compliance, and add new oil and gas reserves.

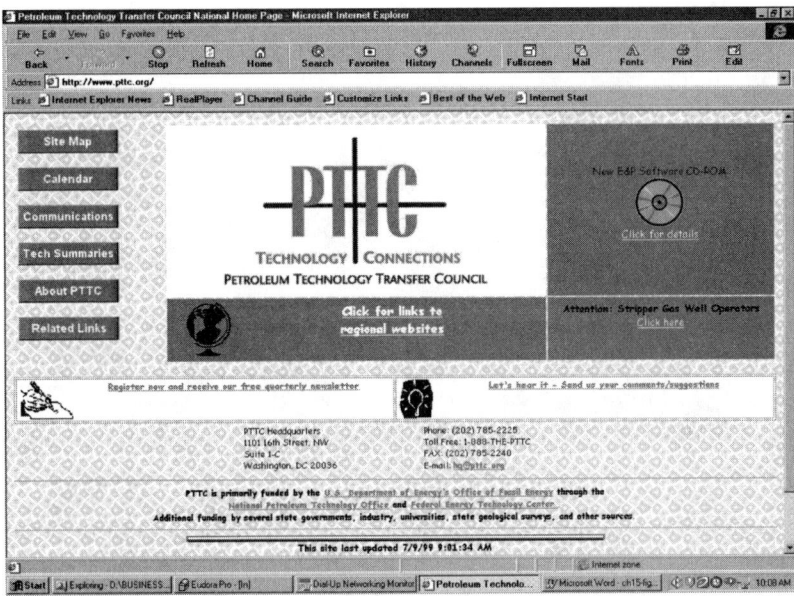

Fig. 15-17 • Petroleum Technology Transfer Council (PTTC)

Chapter SIXTEEN

Case Study 1–
NEWLY DISCOVERED
FIELD DEVELOPMENT PLAN

INTRODUCTION

This chapter presents an example of a development plan for a newly discovered field. It illustrates the application of the reservoir management process/methodology and the role of computer software in making an economically viable development plan for the field.

Offshore Wizard Field, analogous to many Gulf Coast reservoirs, was recently discovered by rank wildcat Well No. 1, and Wells 2, 3, 4, and 5 were drilled to delineate the field (Figs. 16-1 and 16-2).[1] A drill stem test was performed at the discovery well, and the results indicated two productive zones, 4,000 ft and 4,500 ft sands, with 875 STBOPD and 1,456 STBOPD production, respectively. Well No. 4 was a dry hole, and Well No. 3 penetrated the oil-water contact.

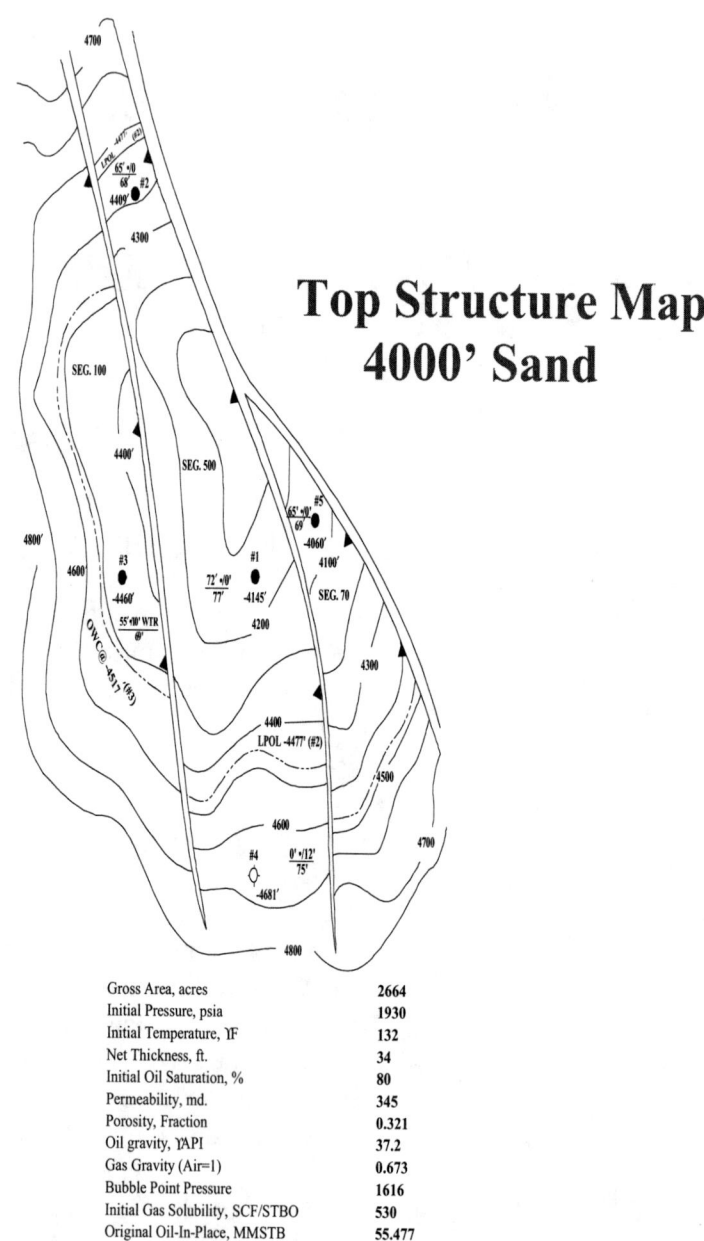

Top Structure Map
4000' Sand

Gross Area, acres	2664
Initial Pressure, psia	1930
Initial Temperature, °F	132
Net Thickness, ft.	34
Initial Oil Saturation, %	80
Permeability, md.	345
Porosity, Fraction	0.321
Oil gravity, °API	37.2
Gas Gravity (Air=1)	0.673
Bubble Point Pressure	1616
Initial Gas Solubility, SCF/STBO	530
Original Oil-In-Place, MMSTB	55.477

Fig. 16-1 • 4000 ft. Sand Structure Map

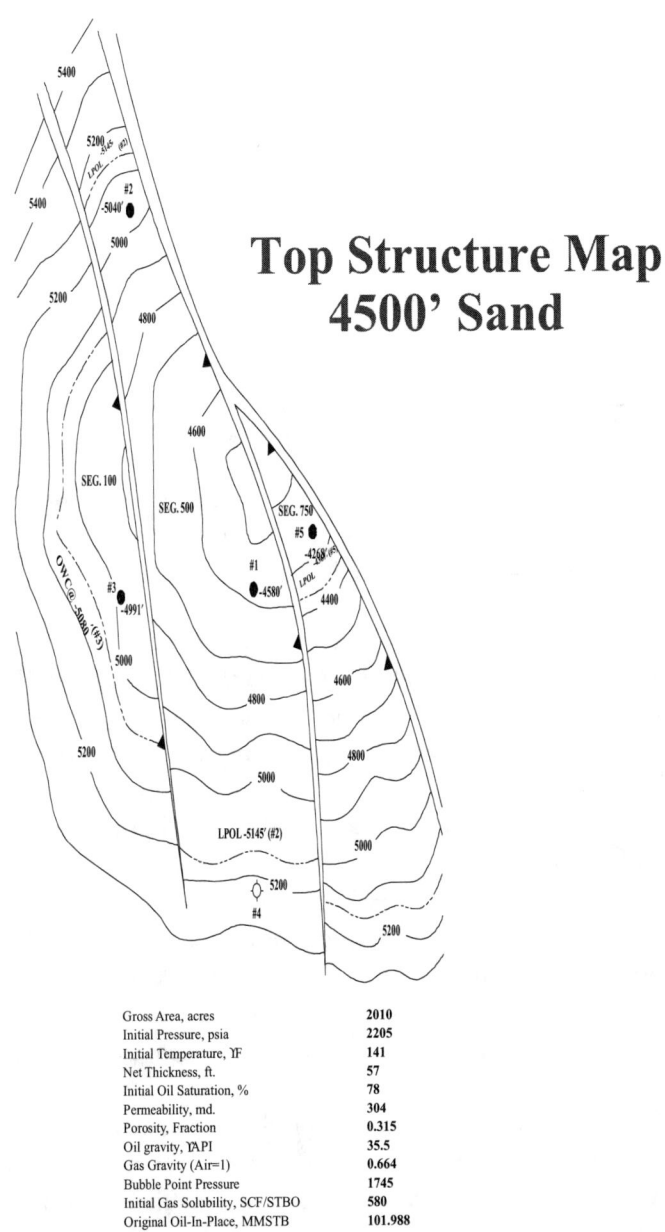

Top Structure Map
4500' Sand

Gross Area, acres	2010
Initial Pressure, psia	2205
Initial Temperature, °F	141
Net Thickness, ft.	57
Initial Oil Saturation, %	78
Permeability, md.	304
Porosity, Fraction	0.315
Oil gravity, °API	35.5
Gas Gravity (Air=1)	0.664
Bubble Point Pressure	1745
Initial Gas Solubility, SCF/STBO	580
Original Oil-In-Place, MMSTB	101.988

Fig. 16-2 • 4500 ft. Sand Structure Map

The available data indicated that this field has a significant amount of potential reserves. An integrated team consisting of geologists, geophysicists, petrophysicists, and structural, reservoir, drilling, completion, and equipment engineers, and other professionals was charged by the management with developing an economically viable plan for the reservoir.

DEVELOPMENT AND DEPLETION STRATEGY

The input of all disciplines, mutual understanding, and interdisciplinary communication were the key to successfully developing an optimum plan. The team needed to address the following main questions in order to come up with an economically viable development and depletion strategy:

1. Recovery scheme—natural depletion or natural depletion augmented by fluid (water or gas) injection?
2. Well spacing—number of wells, platforms, reserves, and economics?

Preliminary data indicated that the reservoir is undersaturated, having the initial pressure several hundred psi above the bubble point. The regional geological data and production experience in this area suggested moderate natural water drive as a potential recovery mechanism in addition to rock and fluid expansion and solution gas drive. However, the possibility of secondary gas cap drive may exist, because of relatively thick pays with high porosity and permeability.

Production from the reservoir by primary depletion and by waterflooding was considered. It was also decided that all the wells would be completed in the lower sands, with the plug back potential in the upper sands when the lower sands become depleted. It was recognized that selective perforation intervals in both sands would maximize the oil recovery and prevent early high GOR production.

In order to realistically forecast oil production rates and reserves, a full-field reservoir simulation study was necessary. Considering several well spacings for the field development, the simulated production performance results were used to economically optimize the number of wells and platforms.

RESERVOIR DATA

The Wizard Field is characterized by faulted structural traps (see structure maps in Figs. 16-1 and 16-2). It consists of two main oil-bearing, highly porous and permeable sandstone formations. The upper 4,000 ft sand is separated by some 500 ft of shale from the lower 4,500 ft sand and both of the formations are intersected by sealing faults. The formations consist of interbedded shales and sands, however, the shales do not appear to be continuous. Based upon permeability variation, the upper and lower sands could be subdivided into two and three layers, respectively. Pertinent reservoir data, which were considered to be reliable, are listed below:

- Figure 16-1: Structure map and reservoir data for 4,000 ft sand
- Figure 16-2: Structure map and reservoir data for 4,500 ft sand
- Table 16-1: Data and sources
- Table 16-2: Layer data

DATA	SOURCES
Structure and Isopach Maps	Seismic Surveys, Revised WithWell Log Information
Reservoir Pressure and Temperature	Drill Steam Test on Well #1
Porosity	Well Logs (Sonic, FDC-CNL, GR, ILD, etc.) from Wells 1, 2, 3, 4, and 5, and Conventional Cores from Well #2
Permeability	Conventional Cores from Well #2
Fluid Saturations	Well Logs from Wells 1, 2,

Table 16-1 • Wizard Field Data and Sources

These data could be stored in a comprehensive geoscience-engineering database for future use.

RESERVOIR MODELING

Considering development of the field using 40, 80, 120, and 160-acre well spacings, a full-field reservoir simulation model was constructed to predict depletion drive performance. A commercial black oil simulator was used with grid cells in 30 columns and 23 rows, oriented along the faults. The properties for both sands were obtained from data given in Table 16-2. The reservoir layers were considered continuous and homogeneous throughout the field. It should be pointed out that the foundation for the field development study is the reservoir description and the reservoir engineer is heavily reliant on this data. The vertical to horizontal permeability ratio was chosen to be 1 to 10 based upon the available conventional core data. PVT and relative permeability data were based upon correlations, using the appropriate parameter values.

Most wells would be initially completed in the lower sands, except for the few wells that would encounter oil-bearing column only in the upper

Reservoir	Layer	Sand Top S.S. ft	Net. Thick. Ft	Porosity %	Permeability md
4000' Sand	A	4351	12	32.2	298
	B	4384	22	32.1	393
4500' Sand	A	4822	13	31.7	221
	B	4854	24	31.5	288
	C	4911	20	31.4	403

Table 16-2 • Wizard Field Layer Data

Primary Depletion

Case	Initial Prod. Rate STBOPD 4000'sd	4500'sd	No. of Platforms	No. of Wells Pre- Prod.	During Prod.	Drilling & Completion Time Days/Well
40-Ac	1000	1750	2	20	35	39
80-Ac	1250	2000	1	10	20	41
120-Ac	1500	2500	1	8	12	45
160-Ac	1500	2500	1	8	7	45

Table 16-3 • Wizard Field Drilling, Completion, and Production Schedule

sands. The bottom two layers in the lower sand and the bottom layer in the top sand would be perforated. The optimum production rates were derived from a Nodal Analysis. Table 16-3 shows the schedule of wells to be brought to production for the various well spacings, along with the initial production rates for each sand. The model used the following well production limitations: economic rate of 30 MSTBOPD, flowing bottom hole pressure of 600 psia, gas-oil ratio of 20,000 SCF/STB, and water cut of 95%.

PRODUCTION RATES AND RESERVES FORECASTS

A 30x23x6-cell grid model was used to predict field-wide primary reservoir performance for 40, 80, 120, and 160-acre well spacings. In addition to the five layers, an impermeable layer was added to isolate the two sands. The simulation runs were made with an active aquifer whose volume is 10 times the reservoir volume. The results show that the larger the spacing, the longer the life, with less oil recovery (Table 16-4 and Fig. 16-3). The production

Parameters	Primary Development				Primary Followed by Waterflood 80-acre
	40-acre	80-acre	120-acre	160-acre	
Investment, $MM	325	222	202	162	220
Reserves, MMSTBO	40.3	40.2	38.7	38.0	81.3
Reserves, %OOIP	25.6	25.5	24.6	24.1	51.6
Economic Life, Yrs	9	1 1	15	15	22
Payout, Yrs.	5.1	4.8	4.7	4.7	4.9
Disc. Cash Flow Return on Inv. (DCFROI), %	29.0	38.8	35.8	40.4	42.7
Net Present Value (NPV), $MM	112	161	144	157	309
Present Worth Index (PWI)	1.63	2.31	2.15	2.49	3.64
Development Costs, $/BO	5.95	3.91	3.62	2.87	2.18

Table 16-4 • Wizard Field Economic Evaluation

performances are shown in Figures 16-4 and 16-5 for the 160-acre well spacing, which turned out to be the optimum spacing case for developing the field, as discussed below.

A sensitivity analysis of the aquifer size on the recovery was made using the 160-acre well spacing. Results, which are shown in Figure 16-6, indicate significant recovery without any aquifer support. The oil recovery is 33.816 MM STBO (21.5% OOIP) for no aquifer influx, as compared to 38.029 MM STBO (24.2% OOIP) with a 10:1 aquifer size.

An additional run was made using 160-acre well spacing for initiating

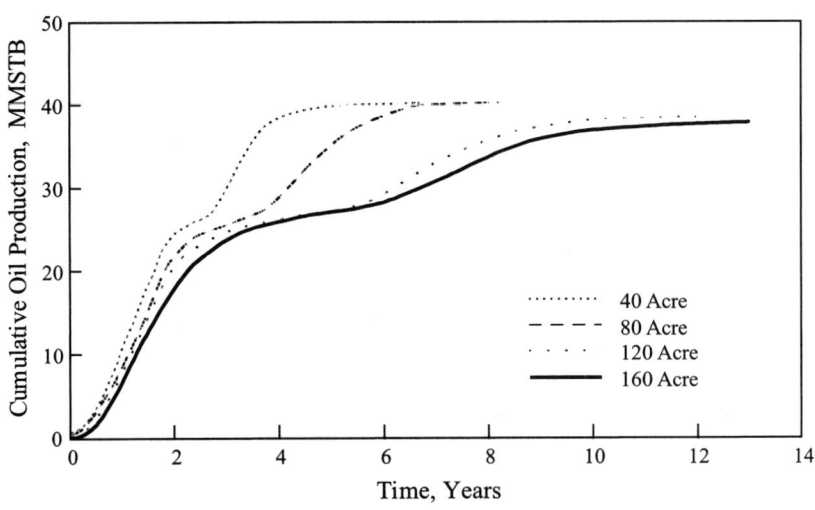

Fig. 16-3 • Effect of Well Spacing on Recovery

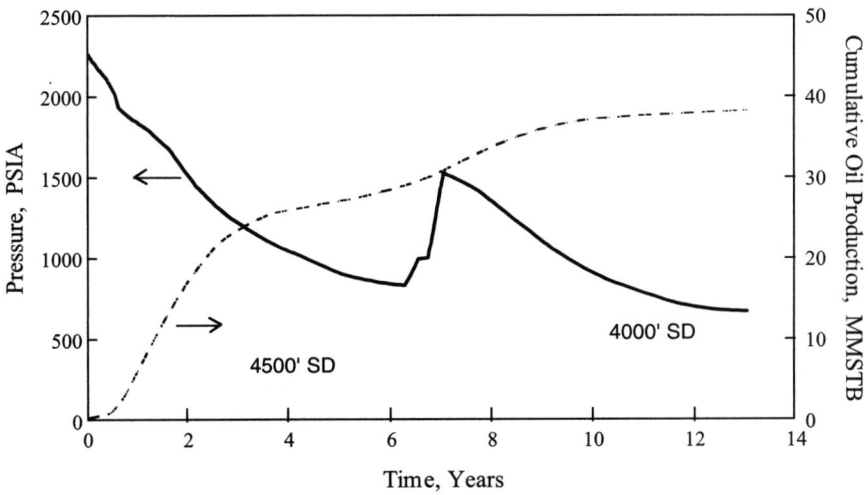

Fig. 16-4 • Pressure and Cumulative Production vs. Time

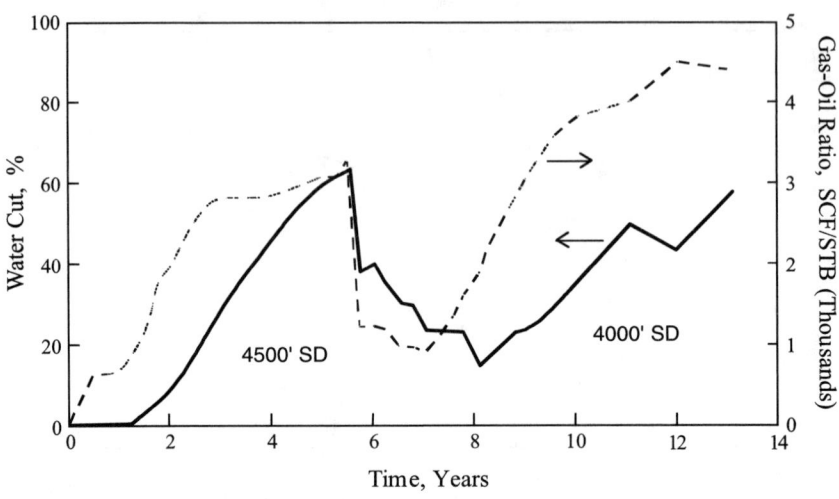

Fig. 16-5 • Effect of Water Cut and GOR vs. Time

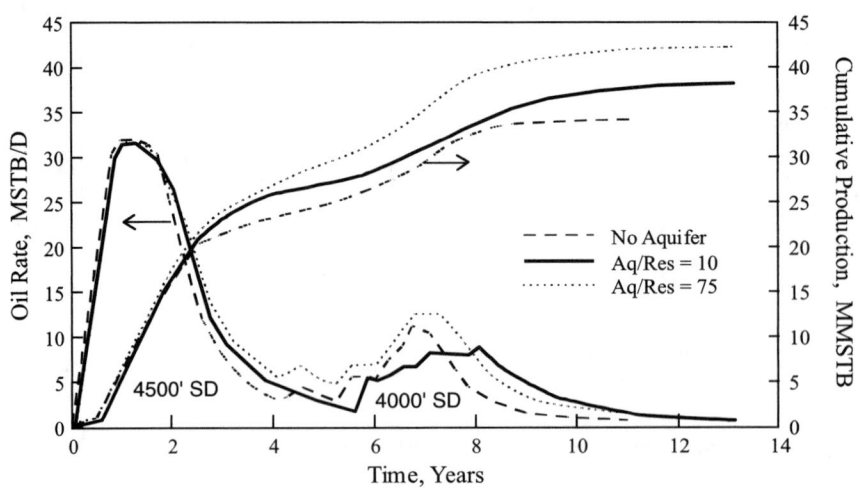

Fig. 16-6 • Effect of Aquifer Size on Rate and Cumulative Production

primary depletion, followed by 80-acre, 5-spot infill waterflooding after two years. In this case, 12 water injection, 18 production, and 3 water source wells would be needed. Water injection would be initiated at the 4,500 ft sand and recompleted to the 4,000 ft sand as all production wells plugged back to this sand. The production performance of this case is compared in Figure 16-7 with the 160-acre depletion case and 10:1 aquifer size. The oil recovery from the waterflood operation was computed to be 81.305 MMST-BO (51.7% OOIP), more than doubling the primary recovery.

FACILITIES PLANNING

The simulated production performance results were used to size platforms, production decks, surface facilities, etc. Also, drilling, well completions, and production practices requirements were established. Subsequently, estimates of capital requirements and operating expenses were made for economic analyses.

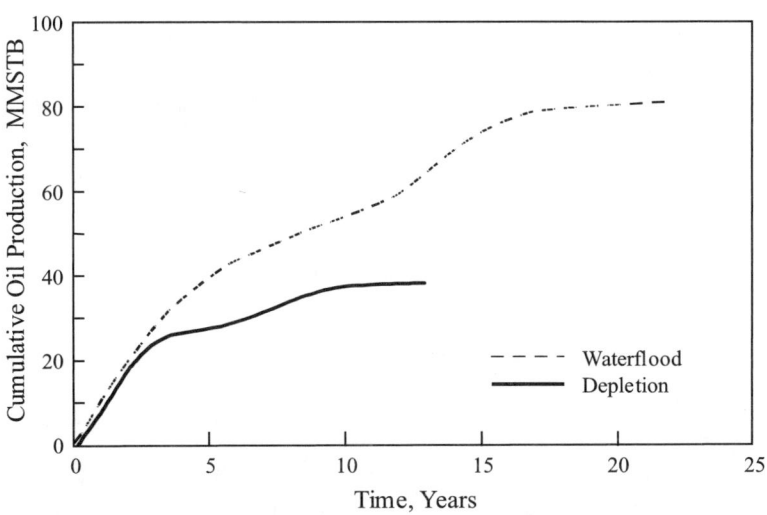

Fig. 16-7 • Cumulative Oil Production vs. Time

ECONOMIC OPTIMIZATION

Using estimated production, capital, operating expenses, and other financial data (Table 16-4), economic analyses of the primary development plans with 40, 80, 120, and 160-acre well spacings were made. Table 16-4 shows the evaluation results for these cases. The 160-acre well spacing case, with the lowest capital investment, development cost, and payout time and the highest PWI, DCFROI, and next highest NPV, offered the economically optimum primary development plan. Even though the 80-acre case yielded the highest NPV ($161 million), the additional capital investment of $60 million over the 160-acre case ($222 million) gave an incremental NPV return of only $4 million.

The 160-acre development case without any aquifer support showed project life of 13 years, DCFROI of 39%, NPV of $140 million, and development costs per barrel of $3.12. Therefore, the 160-acre primary development still looked economically very attractive.

Results of the economic analysis of the waterflood case (Table 16-4) show the highest oil reserves, DCFROI, NPV, PWI, and lowest development costs per barrel of oil. Therefore, the early waterflood offers the most economic means to exploit this field. The platform needs to be designed so the water injection facilities could be installed later, *i.e.*, some deck space would be left for future water injection equipment. Based on the economic evaluation results, the team recommended to its management the initial 160-acre primary development followed by 80-acre, 5-spot infill waterflooding after two years.

IMPLEMENTATION

After management approval of the project, the next major assignment would be to implement the development plan in order to get the production on stream as soon as possible. A project manager with full authority would be needed to manage the various activities including installation of platform and surface facilities, drilling and completion programs, acquiring and analyzing necessary logging, coring, and well test data to better define reservoir characterization. The reservoir database needs to be upgraded, and simulation runs have to be made with the latest data for upgrading the depletion strategy and predicting reservoir performance.

MONITORING, SURVEILLANCE, AND EVALUATION

An integrated and comprehensive monitoring and surveillance program needs to be initiated at the start of production from the field. Dedicated and coordinated efforts of the various functional groups working on the project are essential. The production performance of the reservoir needs to be monitored and reviewed periodically to ensure the development plan is working. This can be done effectively by comparing the actual well/reservoir production and pressure behavior with the simulated performance.

REFERENCES

1. Satter, A., Varnon, J. E., and Hoang, M. T.: "Reservoir Management: Technical Perspective," SPE Paper 22350, SPE International Meeting on Petroleum Engineering, Beijing, China, March 24-27, 1992

Chapter *SEVENTEEN*

Case Study 2–
MATURE FIELD STUDY: NORTH APOI/FUNIWA

INTRODUCTION

The mature North Apoi/Funiwa
Field (Fig. 17-1), operated by Texaco
Overseas (Nigeria) Petroleum Company
Unlimited (TOPCON), is located offshore Nigeria (Fig. 17-2). North Apoi was
discovered in 1973, followed by Funiwa,
an extension of North Apoi, in 1978.
The fields consist of north-west/south-
east tending anticline with several major
faults (Figs. 17-3 and 17-4), and multiple
sand reservoirs at depths between 4,900 ft
(ss) and 7,000 ft (ss) (Fig. 17-5). Figure
17-6 shows typical log and core analysis
data of North Apoi Well #34. Basic reservoir data for Ewinti and Ala Series are
shown in Table 17-1. Sixty-four wells,
including 58 commercial wells, were
drilled as of June 30, 1995. Primary producing mechanisms are a combination of

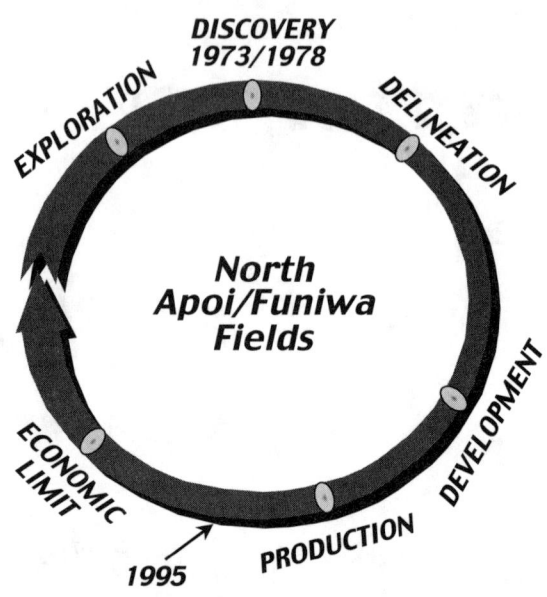

Fig. 17-1 • North Apoi/Funiwa Life Cycle

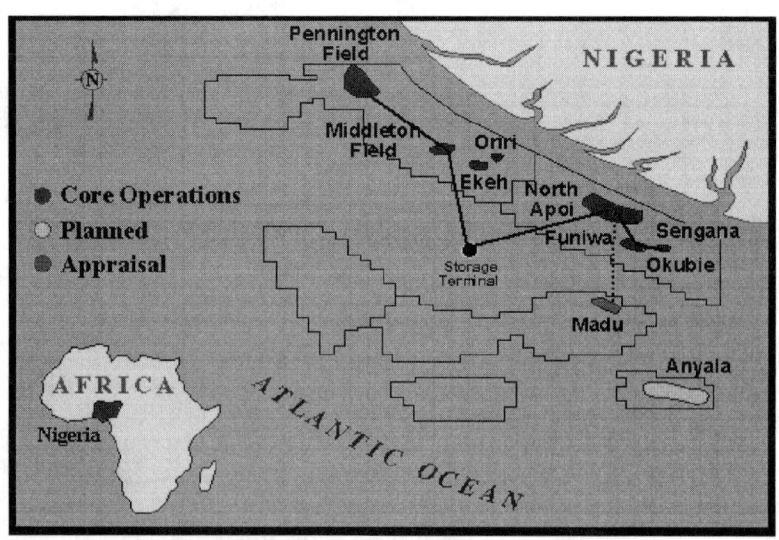

Fig. 17-2 • North Apoi/Funiwa Field Location Map

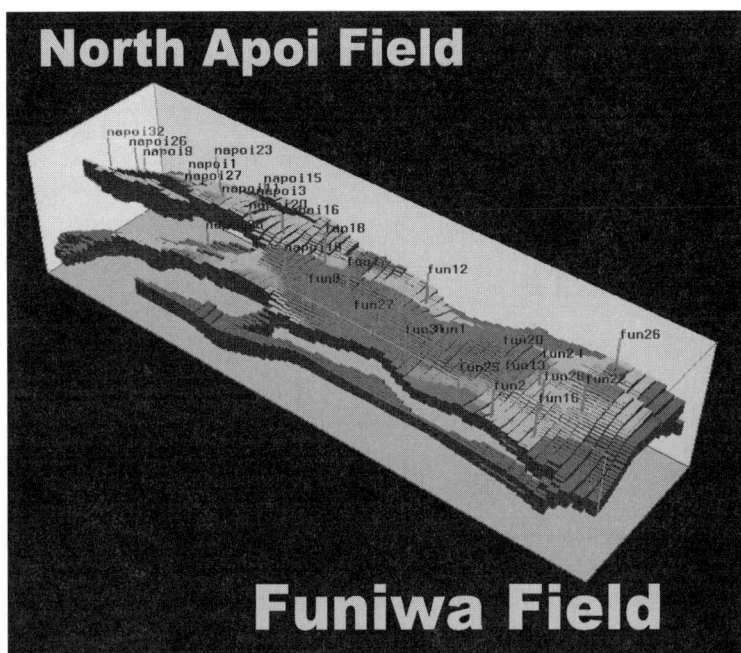

Fig. 17-3 • North Apoi/Funiwa Field

depletion, gas cap, and natural water drives. Ewinti-5, -6, -7 and Ala-3, -5, -7 are the major producing sands.[1]

An integrated team of geoscientists and engineers from TOPCON, Nigerian government organizations, and Texaco E & P Technology Department (EPTD) were charged to review the field's six largest reservoirs in three phases to evaluate and capture the upside potential of the reserves.

The first phase study involved the Ewinti-5 and Ala-3 reservoirs, which contain more than 50% of the field's booked reserves (Fig. 17-7). This work was initiated in March and completed in June, 1995 at EPTD facilities in Houston. The second phase study, covering the Ewinti-7 and Ala-5 reservoirs, and the third phase study, covering Ewinti-6 and Ala-7 were also carried out in Houston, taking three months each.

These integrated studies exemplify the close alignment between EPTD and TOPCON in the transfer and application of leading edge technologies

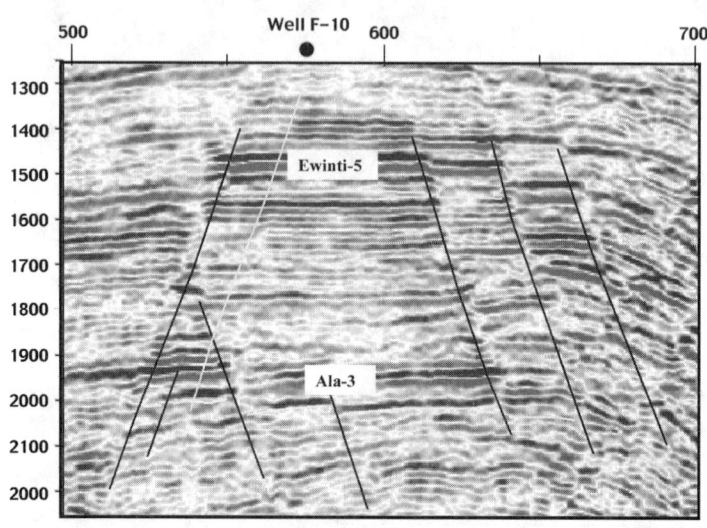

Fig. 17-4 • North Apoi/Funiwa Field Seismic Line

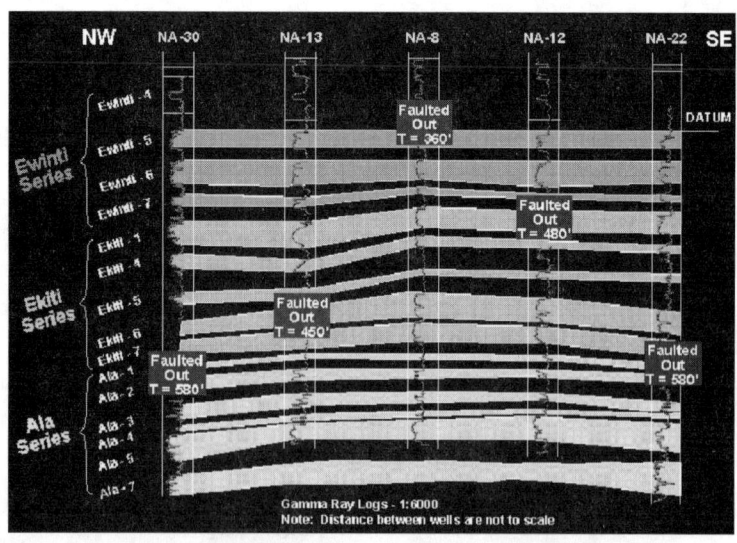

Fig. 17-5 • North Apoi/Funiwa Stratigraphic Cross-Section

in support of TOPCON's growth plan. The results of these mature-field studies are presented here.

OBJECTIVES

The objectives of the studies were to determine:

- ultimate primary recovery
- optimum recovery with additional vertical and horizontal wells, and workovers, including gas lift
- EOR potential

CHALLENGES

Challenges faced in the studies were:

- mature reservoirs
- declining production
- increasing operating cost
- unrealistic recovery factors
- need to enhance asset value

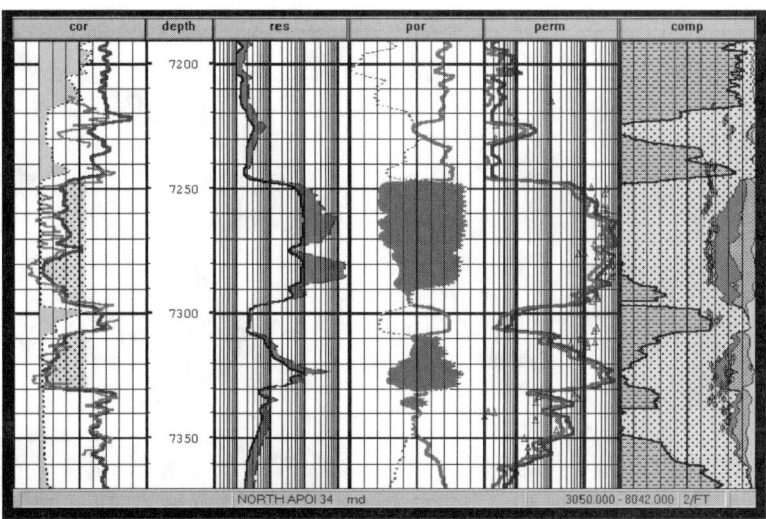

Fig. 17-6 • North Apoi Well #34 Log and Core Data

	EWINTI-5	ALA-3
DEPTH - FT SS	5000	7000
TRAP	STRUCTURE/ FAULTING	STRUCTURE/ FAULTING
ROCK TYPE	UNCONSOLIDATED SAND	UNCONSOLIDATED SAND
GROSS THICKNESS - FT	70-130	50-170
POROSITY - %	30	20-25
PERMEABILITY - MD	1500	500-1500
INITIAL PRESSURE - PSIG	2200	3000
RESERVOIR TEMPERATURE - °F	165	222
INITIAL SOLN. GOR - SCF/STB	364	940
INIT. OIL FORM. VOL. FACTOR	1.2	1.6
OIL VISCOSITY - CP	1.5	0.5
OIL GRAVITY - °API	28	40
GAS GRAVITY (AIR = 1)	0.6	0.7
DRIVE MECHANISM	GAS CAP/ STRONG WATER DRIVE	GAS CAP/ WEAK WATER DRIVE
ORIG. OIL IN PLACE - MMSTB	293	214
CUM. PROD. 12/94 - MMSTB	85	47

Table 17-1 • North Apoi/Funiwa Field General Data

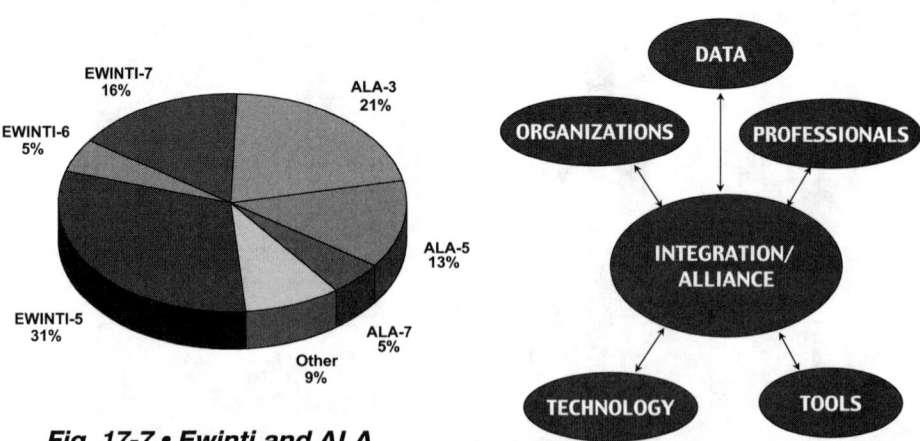

*Fig. 17-7 • Ewinti and ALA
Reserves (as of Dec. 31, 1994)*   *Fig. 17-8 • Integration/Alliance*

APPROACH

The approach taken was:

1. Review geoscience and engineering data
2. Classical material balance analysis
3. Decline curve analysis
4. Reservoir simulation analysis
    a. reservoir description
    b. full-field performance history match
    c. full-field performance prediction
5. Plan strategies and forecast performance
    a. under existing conditions
    b. with workovers and infill wells
    c. with gas lift
    d. with water injection

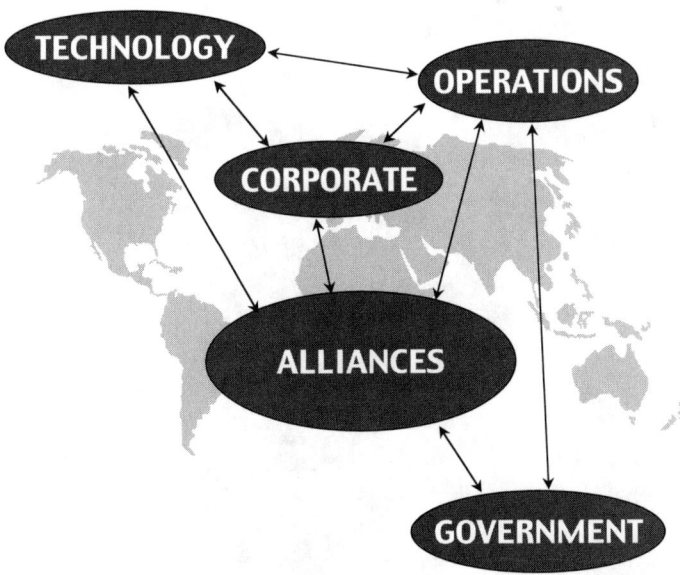

Fig. 17-9 • Organizational Alliance

DELIVERABLES

At the outset of the studies, all parties agreed on the following deliverables:

- Improved reservoir description
- Updated OOIP
- Reserves addition
- Better reservoir management skills and strategies
- Technology transfer/application

The studies utilized integration/alliance (Fig. 17-8) of organizations (Fig. 17-9), data and software (Fig. 17-10), professionals working together as a team (Fig. 17-11 and Table 17-2), tools and technologies (Fig. 17-12). Multidisciplinary data used and their sources are listed in Table 17-3.

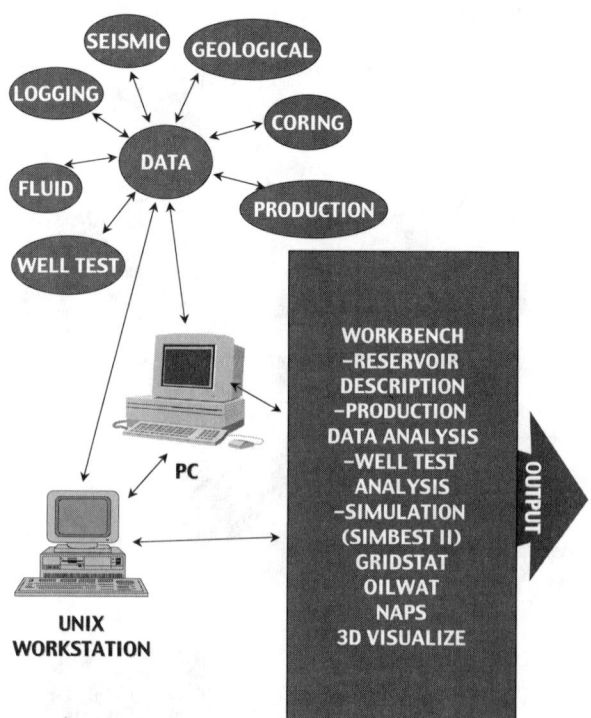

Fig. 17-10 • Data and Software Integration

Integrated geoscience and engineering models were developed using revised maps based upon reprocessed and re-interpreted 3-D seismic survey data of 1986. Well log and core analysis data, rock and fluid properties, well test data and other engineering data, plus 20 years of field production history were also incorporated into the reservoir description.

DECLINE CURVE ANALYSIS

Integrated Petroleum WorkBench software of Scientific Software Intercomp (SSI) consists of reservoir description, well test analysis, production data analysis, and black oil simulation modules. The production data analysis module was used to make decline curve analyses on all wells where it was appropriate, *i.e.*, wells that had established declining production during a period where well conditions had not changed appreciably. The results were compared to the simulation predictions.

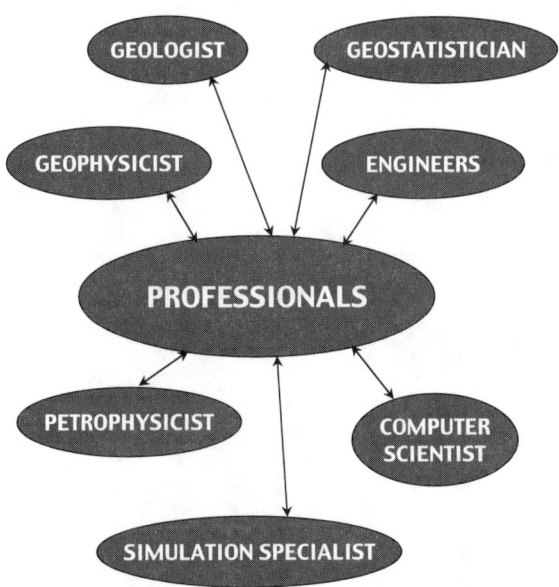

Fig. 17-11 • Integration of Professionals

Team

- **Texaco Nigerian Division**
  - Engineers - 3
  - Geophysicist - 1
  - Geologist -1

- **Nigerian Government Professionals**
  - Engineers - 2

- **Texaco EPTD Specialists**
  - Geophysics -1
  - Computer Support -1
  - Geostatistics -1
  - Horizontal Wells- 2
  - Petrophysics -1
  - Project Management -1
  - Reservoir Engineering - 2
  - 3D Visualization - 2

- **Texaco Management Support and Commitment**

Table 17-2 • Team Members

Data Sources

DATA	SOURCE
STRUCTURE & ISOPACH MAPS	3D SEISMIC & WELL LOGS
POROSITY, PERMEABILITY, & FLUID SATURATIONS	WELL LOGS, CORES, & CORRELATIONS
FLUID CONTACTS & FORMATION TOPS	WELL LOGS
RESERVOIR PRESSURE & TEMPERATURE	WELL TESTS
PVT PROPERTIES	BOTTOM HOLE SAMPLES & CORRELATIONS
RELATIVE PERMEABILITIE S	CORES & CORRELATIONS
PRODUCTION RATES & HISTORY	WELL TEST & ALLOCATION SUMMARY

Table 17-3 • Data Sources

CLASSICAL MATERIAL BALANCE

Classical material balance analysis was considered to be a pre-requisite to reservoir simulation. The EPTD-developed OILWAT material balance software was used for estimating original oil-in-place and primary drive mechanisms. The material balance analysis showed that the primary production mechanism of the Ewinti-5, Ewinti-7 and Ala-5 sands is strong water drive, with additional support from gas cap drive and solution gas drive. The Ala-3 reservoir demonstrates weak water drive plus gas cap drive and solution gas drive.

Original Oil in Place (OOIP) calculations provided comparisons with the simulation results (Fig. 17-13), while the drive mechanisms were used as input to the simulation model descriptions.

RESERVOIR SIMULATION

SSI's black oil simulator was utilized for full-field performance history match and forecasts. The stepwise history matching procedure consisted of

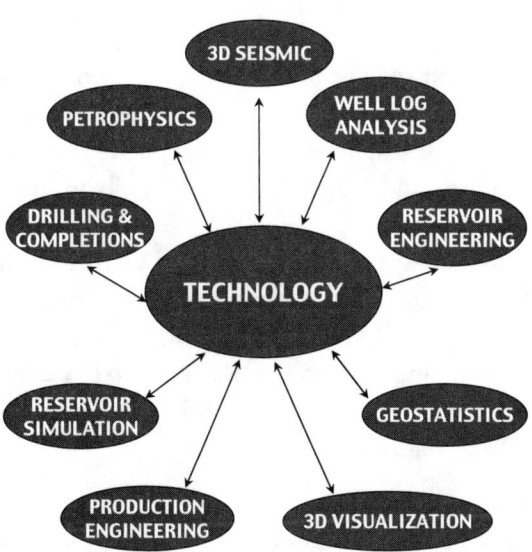

Fig. 17-12 • Integration of Technology

pressure matching followed by saturation matching. Pressure matching was achieved by specifying the historical total three-phase voidages for the wells, while adjusting pore volumes, aquifer strength, and fault connections. Three-dimensional seismic survey data were re-examined for validating the reservoir models. Pressure matching ensured that the reservoirs' historical total (three-phase) voidages were duplicated both for the total reservoir and for each of the wells.

The OOIP values estimated from classical material balance analyses and simulation techniques are comparable to each other but are substantially higher than the previously booked values (Fig. 17-13).

Good history matches using the black oil simulator were achieved for most of the wells by adjusting the usual reservoir parameters within their accepted ranges of uncertainty. Difficulties matching a few wells, however, led to questions about the structure maps. Here, the interaction between the geoscience and engineering members of the team proved very beneficial. The 3-D seismic survey data were re-examined to validate the reservoir model. Ultimately, some areas of poor seismic resolution were re-interpreted, leading to successful history matches in all wells.

After reservoir performance history matching using the black oil simulator, model prediction runs were made under various investment scenarios for optimally draining the reservoirs, including additional take points, horizontal wells, gas lift, and water injection. Opportunities were identified for performing workovers and placing additional wells to improve drainage in the Funiwa area.

Nodal analysis software (NAPS) was used to treat wellbore hydraulics. Computed WorkBench results were imported into a 3-D visualizer (REVIEW) for more comprehensive viewing of the results. Performance forecasts for the remaining period of the production contract were made under different operating scenarios in order to determine the optimum development plan as follows:

Case 1: Primary depletion with the current wells and production limitations (Base Case)

Case 2: Base Case + Infill Wells + Workovers

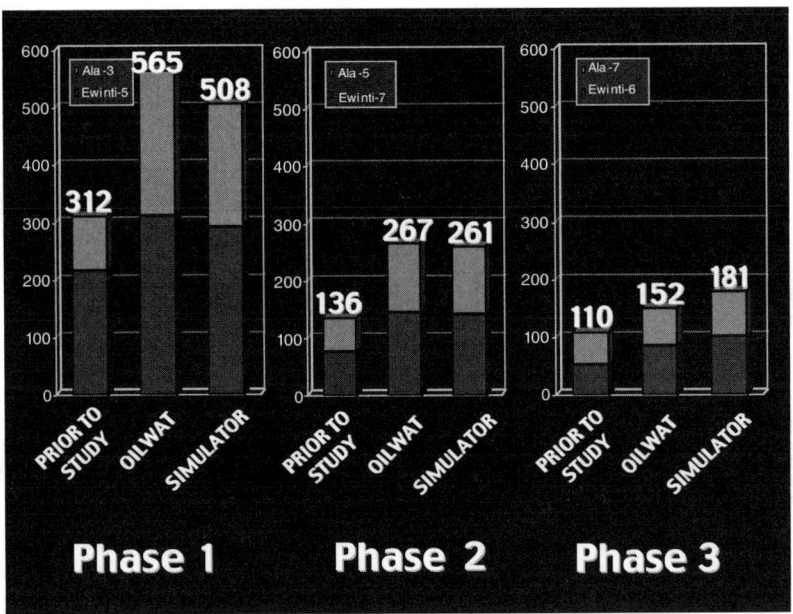

Fig. 17-13 • Original Oil in Place

Case 3: Case 2 + Gas Lift
Case 4: Case 3 + Water Injection, applicable to Ala-3 sand

Since the Ala-3 reservoir has weak natural water drive, the studies showed that recovery could be improved with water injection. The Ewinti-5, Ewinti-7, and Ala-5 reservoirs, on the other hand, have strong water drives, and thus no water injection case was attempted.

The estimated reserve increases from the simulation studies are given in Figure 17-14. The new estimated reserves are substantially more than the previously booked values. Recommendations for the six reservoirs studied in Phases 1, 2, and 3 include 10 horizontal wells, 4 deviated wells, 1 replacement well, and 4 workovers. All infill and workover wells are located in the Funiwa field. Since the current drainage patterns in the North Apoi area are adequate, no additional offtake points are necessary there.

The placement of the wells was determined from the simulator-calculated fluid saturation distributions initially and throughout the producing life of the reservoirs. Figure 17-15, for example, shows the oil saturation distributions in the Ewinti-5 Layer 4 model initially and at the time of the study (1995), plus the predicted distributions at the end of the lease expiration (2008), for the base case and for the infill/workover cases. The locations of the horizontal wells shown in this figure were based upon high remaining base case saturation predicted for 2008.

TOPCON and the Nigerian government acted quickly on the study recommendations. Within nine months from the start of the Phase 1 study, two successful horizontal wells were drilled and completed in the Funiwa Ewinti-5 reservoir (Table 17-4). The first well came on production at 2,670 BOPD of oil from a 700-foot horizontal section. The second well has a 1,600 foot horizontal pay section and produced at 4,020 BOPD.

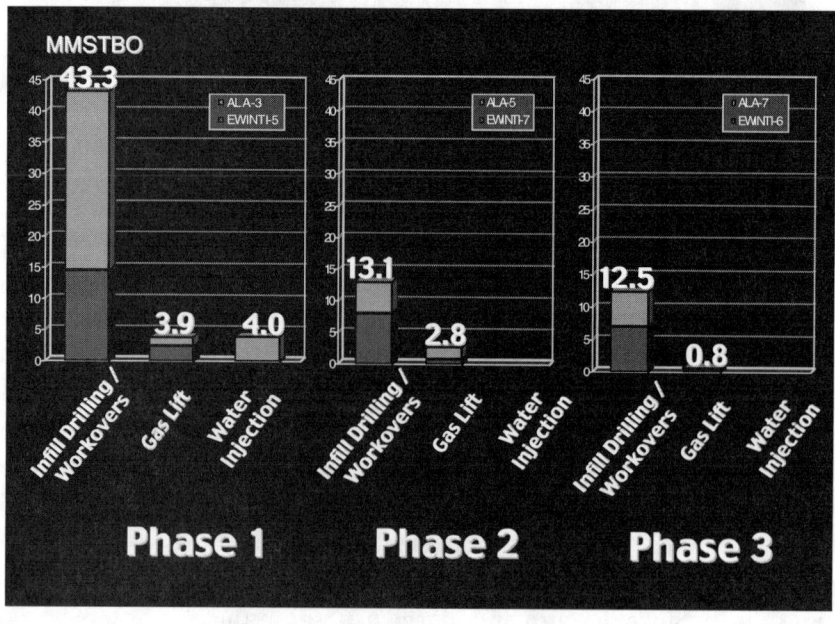

Fig. 17-14 • Reserves Addition Summary

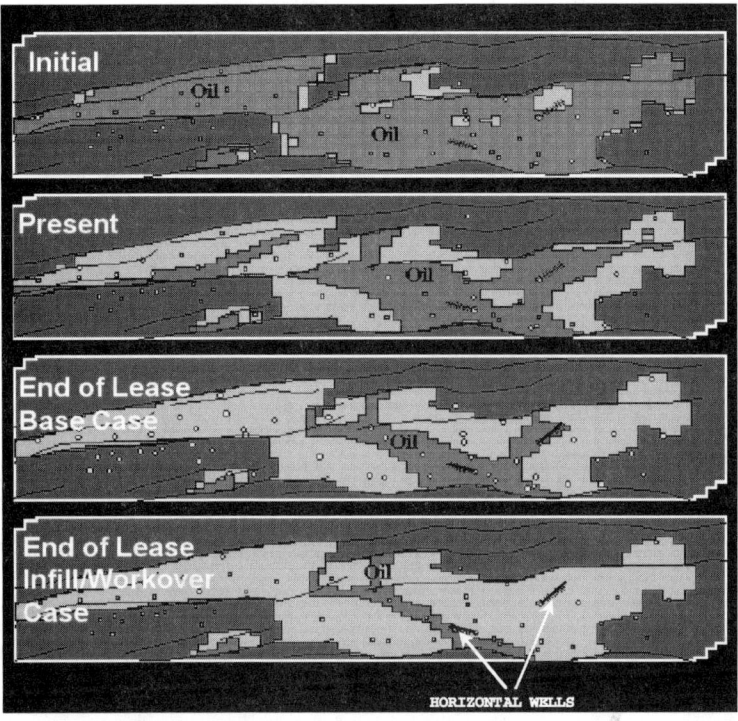

Fig. 17-15 • Oil Saturation Distribution

CONTRIBUTIONS

The role played by each partner in this alliance—operations, technology, government, and vendor—was essential to the successful outcome of the project.

TOPCON, the operator, recognized both the need for the investigation to be made and the benefits of collaboration. Their engineers and geoscientists provided all the field data plus an in-depth knowledge of reservoirs and current producing operations. They performed a majority of technical project work themselves.

EPTD, the technology center, provided project coordination, computer software and hardware, software training, and specialized expertise in 3-D

Horizontal Well Results

	WELL F-32	WELL F-33
ZONE	EW-5	EW-5
HORIZONTAL SECTION (ft)	1600	700
BOPD	4020	2670
BS&W (%)	1	0
GOR	237	260

Table 17-4 • Funiwa Ewinti-5 Reservoir—Horizontal Well Results

seismic interpretation, well log analysis, reservoir simulation, 3-D visualization, and horizontal drilling.

Nigerian government engineers took an active role in project work. Their participation ensured that all regulations would be met and the government's interests were considered early in the planning of proposed operations. This led to rapid approval and early commencement of drilling.

SSI, the WorkBench software vendor, provided consultants who contributed significantly to the timely completion of the projects. They arranged for the availability of extra software licenses for the project and provided technical support for this first major project at EPTD using their WorkBench product.

The joint efforts resulted in significant cycle time reduction and set an excellent example of integration and alliance.

CONCLUSIONS

The conclusions made from the integrated study were:

- 3-D seismic improved reservoir description
- OOIP was revised by more than 50%
- Upside potential of reserves additions is significant
- Recovery factors estimated from 30% to 55%
- Teamwork and integration were critical to project success
- This study sets an example for effective technology transfer and application

TOPCON has since completed several other studies, based upon the success of this one. The reserve additions resulting from these studies are substantial.

REFERENCES

1. Akinlawon, Y., Nwosu, T., Satter, A. and Jespersen, R.: "Integrated Reservoir Management Doubles Nigerian Field Reserves," Hart's Petroleum Engineer International, October 1996

Chapter EIGHTEEN

Case Study 3–

WATERFLOOD PROJECT DEVELOPMENT

INTRODUCTION

A field that was discovered many years ago is now depleted. It consists of a simple domal structure, and five of the nine wells drilled were producers (Fig. 18-1). Primary producing mechanisms were fluid and rock expansion (reservoir pressure above the bubble point), solution gas drive, and limited natural water drive. There is limited data availability. Even the gas, oil, and water production data are somewhat unreliable. Reservoir pressures were not routinely monitored. This is typical when dealing with old, mature fields.

An integrated team of geoscientists and engineers was charged by the management to review past performance and investigate waterflood potential of this field (hypothetical, but akin to real life reservoirs). The team's approach was to:

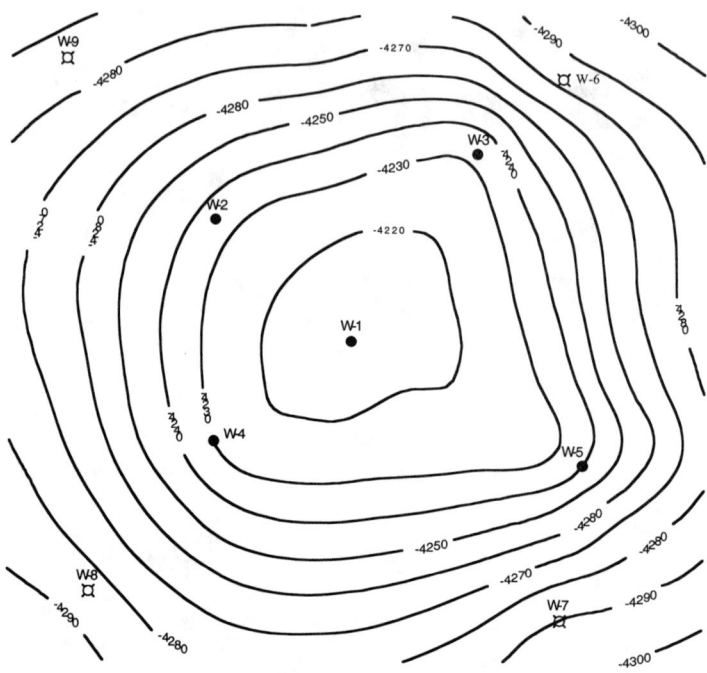

Fig. 18-1 • Top Structure Map

- build an integrated geoscience and engineering model of the reservoir using available data and correlations for fluid PVT and relative permeability
- carry out a conceptual simulation of full-field primary performance without history matching since no historical pressure data or reliable production volumes could be obtained
- forecast performance under peripheral and pattern waterflood

This chapter presents the results of the study. Although the field is not real, much can be learned about how to engineer a waterflood project, even with incomplete data.

RESERVOIR DATA

An analysis of the logs from the nine wells showed that the reservoir heterogeneity could be represented by five producing horizons. Permeability was computed from a correlation of porosity vs. permeability. Permeability in the x and y directions are considered to be the same, *i.e.*, no directional permeability. The vertical to horizontal permeability ratio is assumed to be 0.1. Layer properties are presented in the figures as listed below:

Tables 18-1 through 18-5: Structure tops, gross and net thicknesses, porosities, and permeabilities for Layers 1-5

Figures 18-2 through 18-5: Gross and net thickness, porosity, and permeability of Layer 1 as an example

Figure 18-6: Cross-section

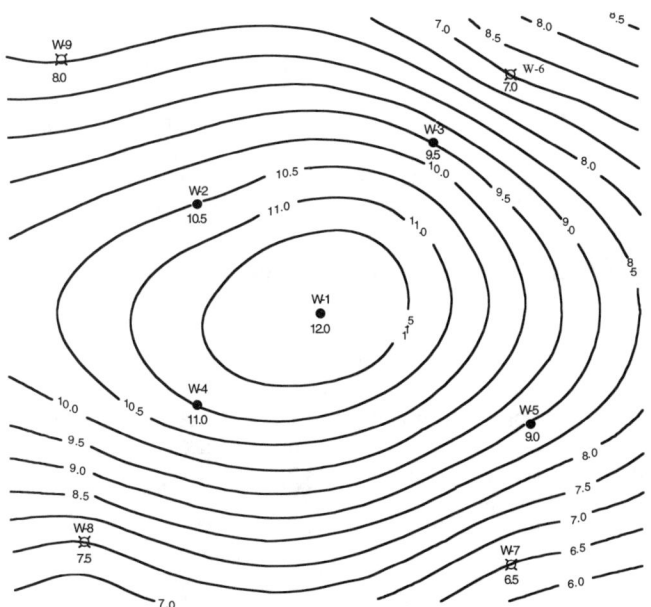

Fig. 18-2 • Gross Thickness—Layer 1

Well No.	HTOP	GROSS (ft)	NET (ft)	PHI (%)	Kx (md)
1	-4216	12.0	12.0	24.0	759
2	-4239	10.5	10.5	21.0	178
3	-4233	9.5	9.5	19.0	68
4	-4230	11.0	11.0	22.0	288
5	-4260	9.0	9.0	18.0	42
6	-4274	7.5	7.5	15.0	10
7	-4285	6.5	6.5	13.0	4
8	-4280	7.0	7.0	14.0	6
9	-4277	8.0	8.0	16.0	16

Table 18-1 • Reservoir Description: Layer No. 1

Well No.	HTOP	GROSS (ft)	NET (ft)	PHI (%)	Kx (md)
1	n/a	16.0	16.0	23.0	68
2	n/a	14.0	14.0	20.0	110
3	n/a	12.5	12.5	18.0	42
4	n/a	14.5	14.5	21.0	178
5	n/a	12.0	12.0	14.0	26
6	n/a	10.0	10.0	14.0	6
7	n/a	8.5	8.5	12.0	2
8	n/a	9.5	9.5	13.0	4
9	n/a	10.5	10.5	15.0	10

Table 18-2 • Reservoir Description: Layer No. 2

Well No.	HTOP	GROSS (ft)	NET (ft)	PHI (%)	Kx (md)
1	n/a	21.0	21.0	22.0	288
2	n/a	18.5	18.5	19.0	68
3	n/a	16.5	16.5	17.0	26
4	n/a	19.5	19.5	20.0	110
5	n/a	15.5	15.5	16.0	16
6	n/a	13.0	13.0	14.0	6
7	n/a	11.5	11.5	12.0	2
8	n/a	12.5	12.5	13.0	4
9	n/a	14.0	14.0	15.0	10

Table 18-3 • Reservoir Description: Layer No. 3

The reservoir contains 33° API crude oil at 2,332 psia original reservoir pressure and 123° F temperature. It is undersaturated, having the initial pressure several hundred psi above the bubble point (1,855 psia). Fluid properties and gas-oil and water-oil relative permeability data that were obtained from correlations, are shown in Figures 18-7 through 18-10.

RESERVOIR MODELING

Using Baker Hughes SSI's black oil simulator with a 25x25x5 grid (3,125 cells, each 1.02 acres), a full-field reservoir simulation model was constructed to predict primary performance. The model used the following well production limitations:

Well No.	HTOP	GROSS (ft)	NET (ft)	PHI (%)	Kx (md)
1	n/a	11.0	11.0	21.0	178
2	n/a	9.5	9.5	18.0	42
3	n/a	8.5	8.5	17.0	26
4	n/a	10.0	10.0	19.0	68
5	n/a	8.5	8.5	16.0	16
6	n/a	7.0	7.0	13.0	4
7	n/a	6.0	6.0	11.0	1
8	n/a	6.5	6.5	12.0	2
9	n/a	7.5	7.5	14.0	6

Table 18-4 • Reservoir Description: Layer No. 4

Well No.	HTOP	GROSS (ft)	NET (ft)	PHI (%)	Kx (md)
1	n/a	13.0	13.0	20.0	110
2	n/a	11.5	11.5	17.0	26
3	n/a	10.5	10.5	16.0	16
4	n/a	12.0	12.0	18.0	42
5	n/a	10.0	10.0	15.0	10
6	n/a	8.0	8.0	12.0	2
7	n/a	7.0	7.0	11.0	1
8	n/a	7.5	7.5	12.0	2
9	n/a	8.5	8.5	13.0	4

Table 18-5 • Reservoir Description: Layer No. 5

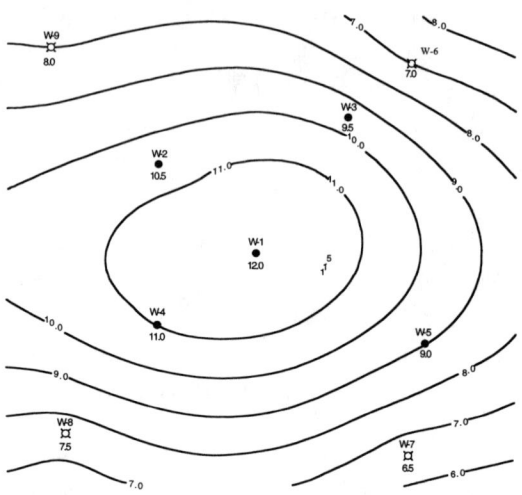

Fig. 18-3 • Net Thickness—Layer 1

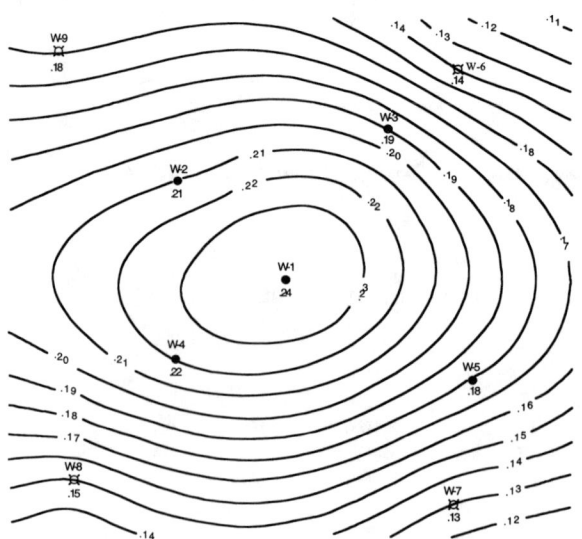

Fig. 18-4 • Porosity—Layer 1

- Economic oil production rate, STB/day   10
- Maximum gas-oil ratio, SCF/STBO   2,500
- Maximum water cut, %   95
- Minimum bottom hole pressure, psia   150
- Completed layers   1,2,3

The original oil-, gas-, and water-in-place were computed to be 26.2 MMSTB, 10.8 BSCF, and 19.1 MMSTB. The reservoir pore volume was 50.1 STB. Primary oil recovery was 4.1 MMSTB (15.7 % OOIP) after 8.3 years. Cumulative oil productions from the individual wells were:

Well	Cumulative Oil Production	Status
1	1,418.0 MSTBO	Producing
2	604.9 MSTBO	Shut-in due to high GOR
3	687.1 MSTBO	Shut-in due to high GOR
4	930.9 MSTBO	Producing
5	467.2 MSTBO	Shut-in due to high GOR

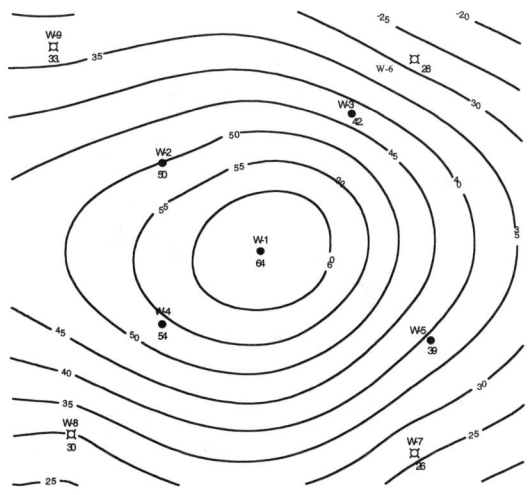

Fig. 18-5 • Permeability—Layer 1

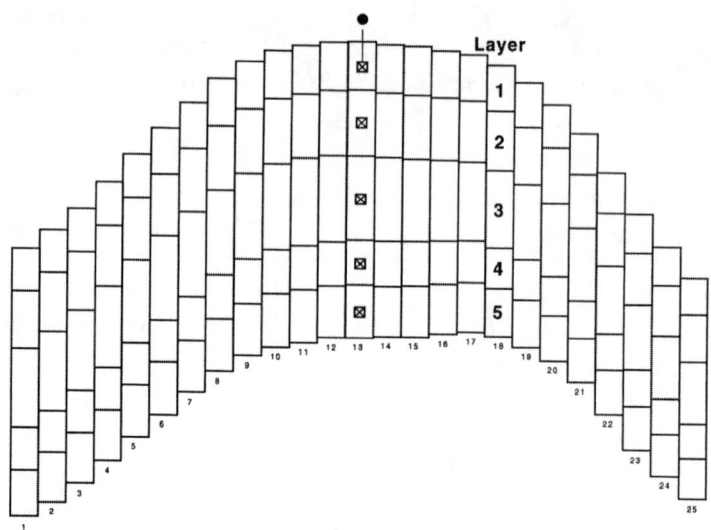

Fig. 18-6 • Cross Section

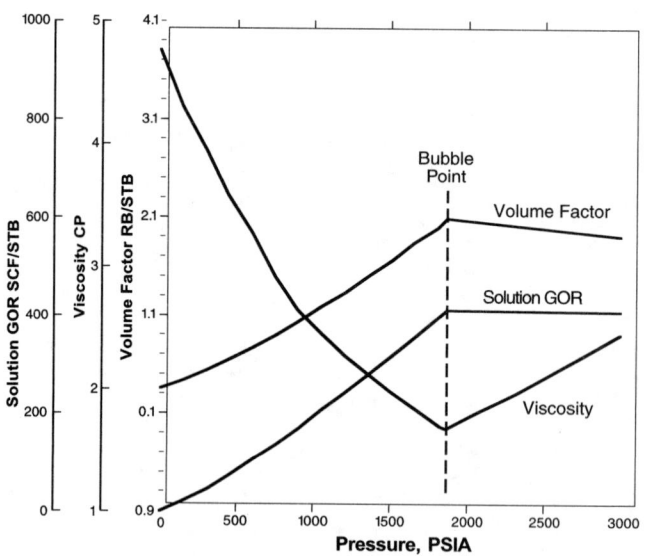

Fig. 18-7 • Oil Properties

Figures 18-11 and 18-12 show field-wide production performance.

PRODUCTION RATES AND RESERVES

Performance forecasts for waterflood recovery were made for the following operating scenarios in order to determine the optimum development plan (Fig. 18-13):

- Case 1 - Peripheral waterflood, using the existing five producing wells and four dry holes at the periphery as injectors
- Case 2 - Enhanced peripheral waterflood, with eight injectors and nine producers
- Case 3 - Single 5-spot pattern waterflood, using the central existing producing well and converting four surrounding wells to injectors

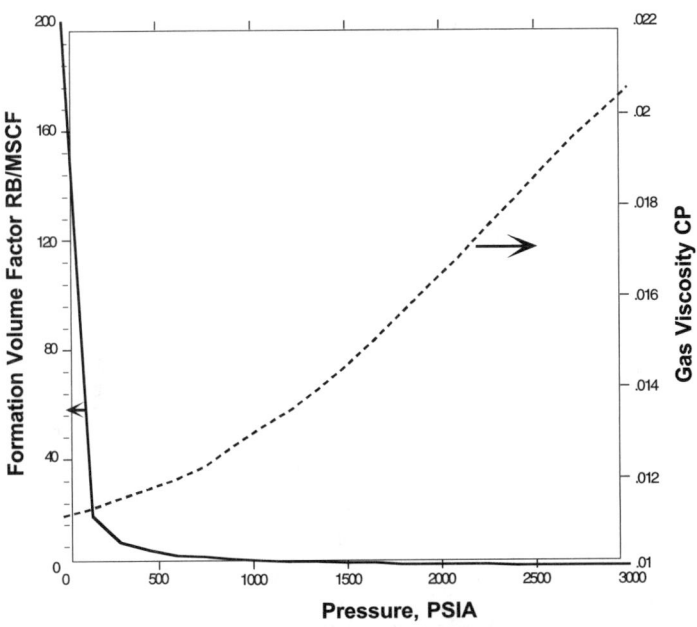

Fig. 18-8 • Gas Properties

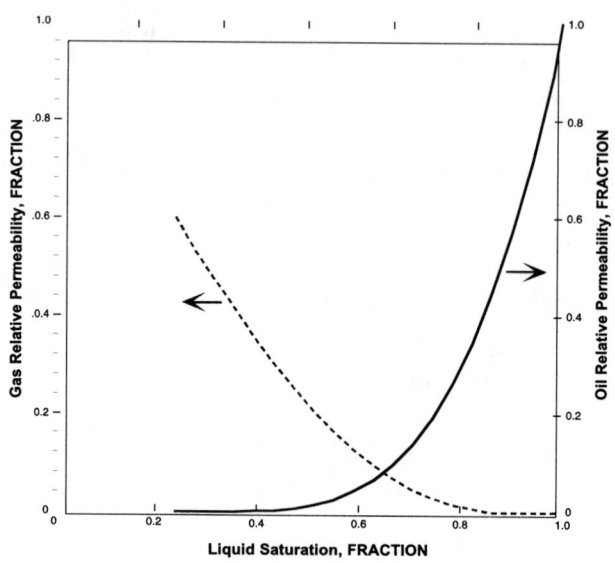

Fig. 18-9 • Gas-Oil Relative Permeability

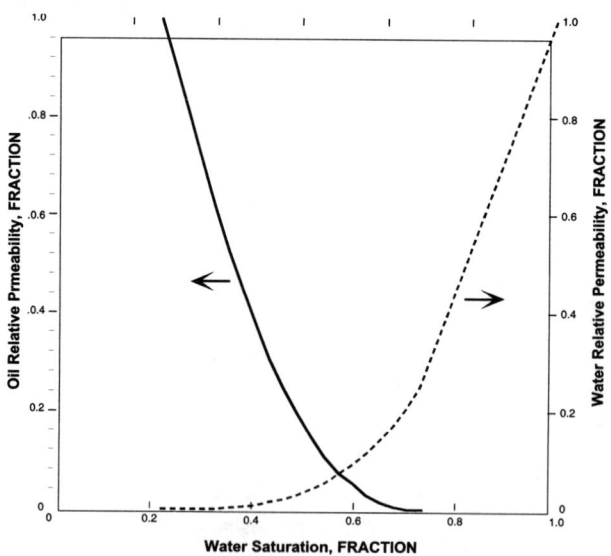

Fig. 18-10 • Water-Oil Relative Permeability

- Case 4 - Four 5-spot pattern waterfloods, using nine injectors and four producers
- Case 5 - Multiple rows of direct line drive waterfloods, using 13 injectors and 12 producers

Calculated results for these cases are presented as listed below:

- Table 18-6: Compares 10 years of annual waterflood oil productions
- Figure 18-14: Compares 10 years of primary and 20 years of waterflood oil productions
- Figure 18-15: Compares waterflood oil recovery vs. pore volume of water injected

Project life being the same, Table 18-6 and Figure 18-14 show significantly higher recoveries in Cases 2 and 5, which have more injectors and producers than the other cases. Case 3, with only one producer shows the least recovery. Or in other words, recoveries are directly related to the number of

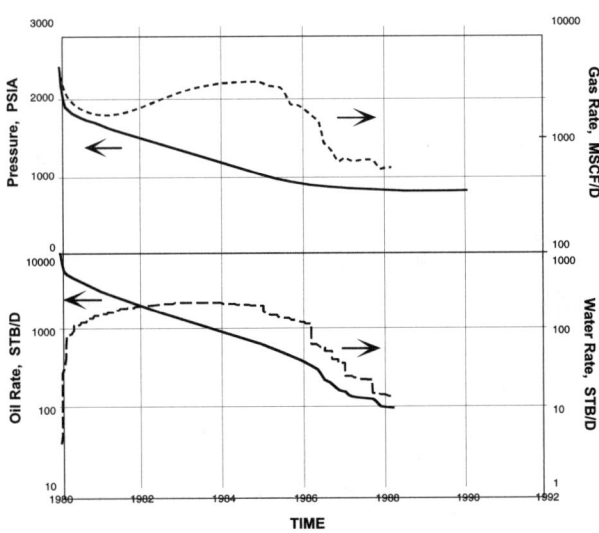

Fig. 18-11 • Fieldwide Primary Production Rates and Pressure Behavior

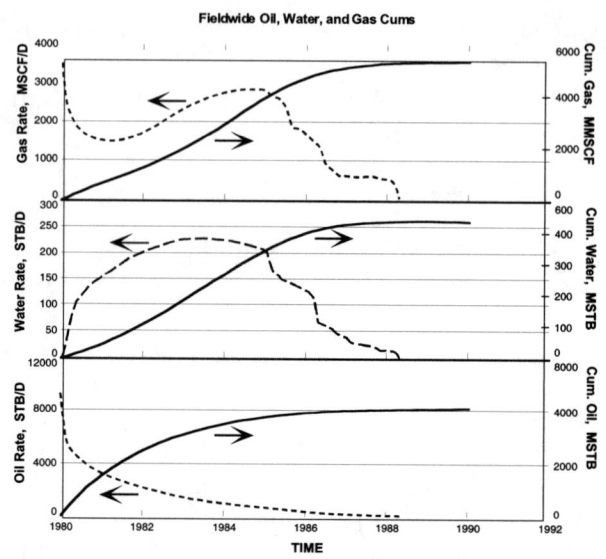

Fig. 18-12 • Fieldwide Primary Cumulative Oil, Water, and Gas Productions

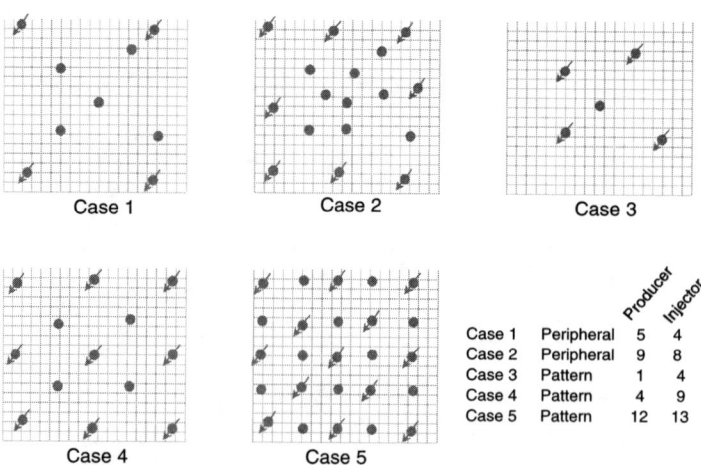

Fig. 18-13 • Development Cases

		Producer	Injector
Case 1	Peripheral	5	4
Case 2	Peripheral	9	8
Case 3	Pattern	1	4
Case 4	Pattern	4	9
Case 5	Pattern	12	13

Year	Case 1	Case 2	Case 3	Case 4	Case 5
1	69.8	275.2	37.5	82.0	279.8
2	49.8	189.7	27.1	127.4	786.2
3	48.7	451.5	27.5	321.0	882.4
4	51.6	656.3	28.3	325.8	605.9
5	55.0	605.7	30.0	293.1	486.8
6	60.8	515.6	35.7	266.5	405.6
7	134.8	460.6	54.0	248.8	341.0
8	234.7	407.6	93.5	237.0	288.9
9	242.1	352.0	155.5	228.1	253.7
10	241.4	305.1	221.1	221.3	220.3
Total	996.0	4,219.5	710.2	2,351.3	4,550.6

Table 18-6 • Produced Oil Volumes

production wells. Figure 18-15 shows that Case 5 is the least efficient recovery scheme when injected water requirements are considered.

More oil recovery with more injection and production wells may not provide the best economically viable scheme. That can only be determined by economic analysis of the cases considered. Also, the limited cases considered here do not necessarily give the answer to best develop the field for waterflooding. These examples simply demonstrate that more than one way of developing the field needs to be investigated in order to determine the potentially most economically viable project.

ECONOMIC EVALUATION

Making a sound business decision requires that the project will be economically viable. That is, it will generate profits satisfying the economic yardsticks of the company. An outline of the procedure for economic evaluation is given in Figure 18-16.[1]

Economic optimization is the ultimate goal of sound reservoir management. It involves more than one scenario or alternative approach to picking the best solution. For example, possible choices and questions concerning

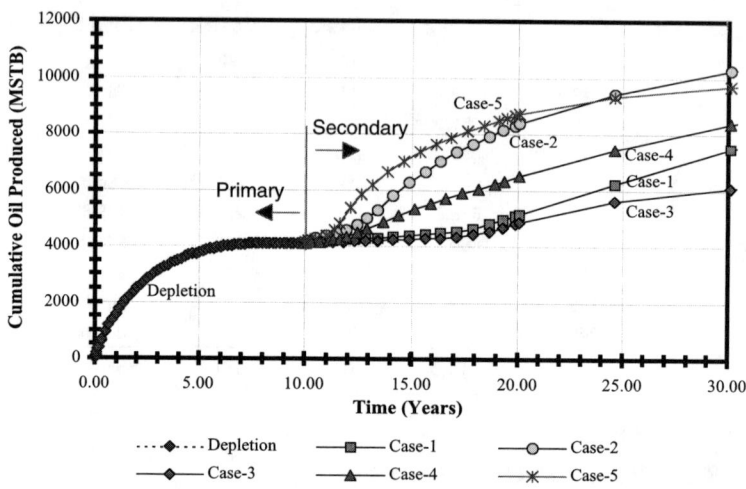

Fig. 18-14 • Cumulative Oil Recovery vs. Time

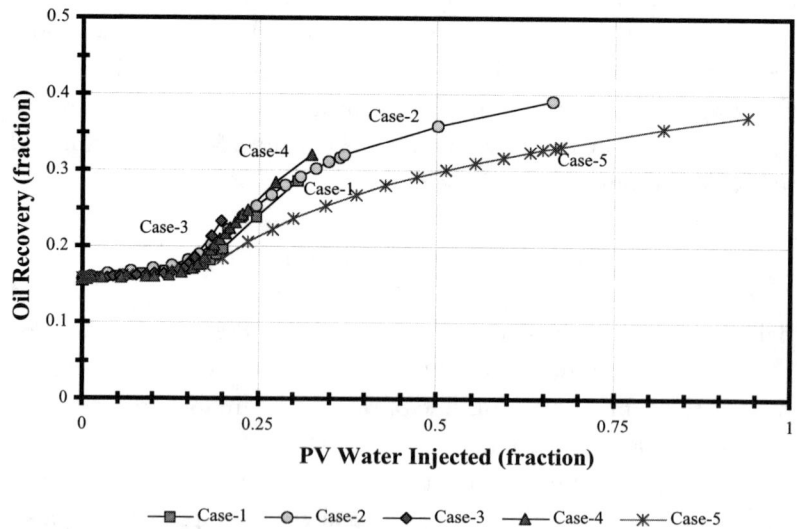

Fig. 18-15 • Oil Recovery vs. Water Injected

the recovery scheme and development plan for a waterflood project:

1. Peripheral vs. inside pattern flood
2. Well spacing - number of wells

The economic analyses and comparisons of the results of the various choices can provide the answer to making the best business decision to maximize profits.

Yardsticks for measuring the value of investments and financial opportunities include:

- *Payout Time*: The time needed to recover the investment. *i.e.*, the time when the undiscounted or discounted cash flow (CF = revenue - capital investment - operating expenses) is equal to zero
- *Profit to Investment Ratio*: The total undiscounted cash flow without capital investment, divided by the total investment
- *Present Worth Net Profit (PWNP)*: The present value of the entire cash flow discounted at a specified discount rate
- *Investment Efficiency or Present Worth Index or Profitability Index*

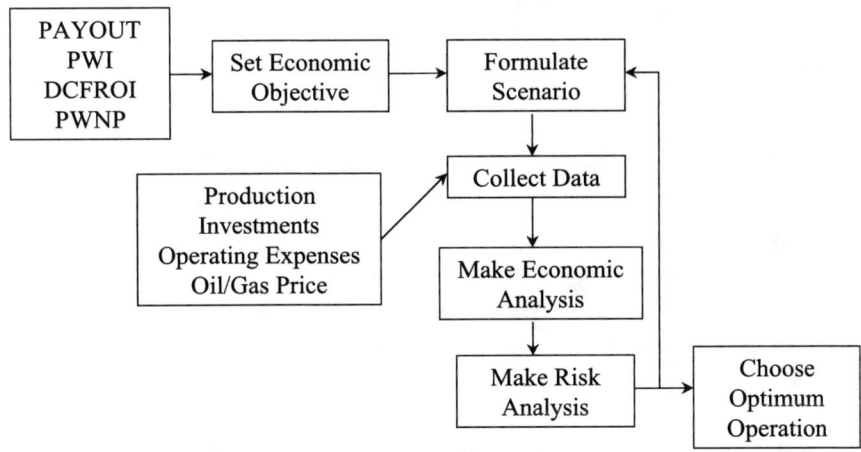

Fig. 18-16 • Economic Optimization

Data	Source/Comment
Oil and gas production rates vs. time	Reservoir engineers/ Unique to each project
Oil and gas prices	Finance and economic professionals/ Strategic planning interpretation
Capital investment (tangible intangible) and operating costs	Facilities, operations and engineering professionals/ Unique to each project
Royalty/production sharing	Unique to each project
Discount and inflation rates	Finance and economics professionals/ Strategic planning interpretation
State and local taxes (production, severance, ad valorem, etc.)	Accountants
Federal income taxes, depletion and amortization schedules	Accountants

Table 18-7 • Economic Data

  (PI): The total discounted cash flow divided by the total discounted investment
- *Discounted Cash Flow Return on Investment (DCFROI)* or *Internal Rate of Return*: The maximum discount rate that needs to be charged for the investment capital to produce a break-even venture. *i.e.,* the discount rate at which the present worth net profit is equal to zero

The data required for economic analysis can be generally classified as production, injection, investment and operating costs, financial, and economic data. Table 18-7 provides a list of pertinent data.

The five waterflood design cases were analyzed to determine the potentially most economically viable project. The procedure used for economic calculation before federal income tax (BFIT) is outlined below:

1. Calculate annual revenues using oil and gas sales from production and unit sales prices
2. Calculate year-by-year total costs including capital investments (drilling, completion, facilities, and abandonment, etc.), operating expenses and production taxes
3. Calculate annual undiscounted cash flow by subtracting total costs from the total revenues
4. Calculate annual discounted cash flow by multiplying the undiscounted cash flow by the discount factor at a specified discount rate

The computational procedure is illustrated in a spreadsheet calculation for Case 2 in Table 18-8.

The results of the economic analysis for the five waterflood cases are presented in Table 18-9, which shows that all the cases are very favorable for waterflooding. Note that federal income taxes are not taken into account.

Case 3 gives the lowest amount of investment, reserves, and development costs, yet has very favorable discounted cash flow return on investment and the highest profit-to-investment-ratio. Case 2 for peripheral flood and Case 5 for pattern flood show the most promise. Case 5 shows the highest present worth net profit; however, this requires 80% more capital than for Case 2, which gives about the same present worth net profit and better discounted cash flow return on investment.

It should be realized that for the Case 2 peripheral flood, recovery estimates might be optimistic, because the reservoir layers were considered to be homogeneous and continuous. This may not represent the real situation. On the other hand, Case 5 for the pattern flood is better suited to treat reservoir heterogeneity and reservoir discontinuity.

The selection of the optimum case will depend on availability of capital, technical considerations, and the risk involved. The sensitivity analysis discussed below shows that Case 2, even if recovery is 20% lower, may still be the best choice.

Economic Evaluation of Case 2

Year	Oil Prod. (MSTB)	Oil Price ($/STB)	Oil Revenue (1)x(2)/1000 ($MM)	Gas Prod. (MMSCF)	Gas Price ($/MSCF)	Gas Revenue (4)x(5)/1000 ($MM)	Total Revenue (3)+(6) ($MM)	Producing Tax (7)xTaxRate ($MM)
1997		19.50	0.00		2.10	0.000	0.000	0.000
1998	275.21	19.50	5.37	1610.5	2.10	3.382	8.749	1.750
1999	189.7	19.50	3.70	503.02	2.10	1.056	4.755	0.951
2000	451.54	19.50	8.81	89	2.10	0.187	8.992	1.798
2001	656.27	19.50	12.80	74.12	2.10	0.156	12.953	2.591
2002	605.71	19.50	11.81	69.86	2.10	0.147	11.958	2.392
2003	515.63	19.50	10.05	60.27	2.10	0.127	10.181	2.036
2004	460.57	19.50	8.98	54.43	2.10	0.114	9.095	1.819
2005	407.65	19.50	7.95	48.8	2.10	0.102	8.052	1.610
2006	351.97	19.50	6.86	42.8	2.10	0.090	6.953	1.391
2007	305.06	19.50	5.95	37.63	2.10	0.079	6.028	1.206
2008	235.47	19.50	4.59	29.77	2.10	0.063	4.654	0.931
2009	227.9	19.50	4.44	28.89	2.10	0.061	4.505	0.901
2010	227.89	19.50	4.44	28.89	2.10	0.061	4.505	0.901
2011	227.9	19.50	4.44	28.89	2.10	0.061	4.505	0.901
Total	5138.47		100.200	2706.67		5.684	105.885	21.177

Year	Capital Investment (9) ($MM)	Discount Factor @21% (10)	Discounted Capital Investment (9)*(10) (11) ($MM)	Operating Cost (12) ($MM)	Total Cost (8)+(9)+(12) (13) ($MM)	Undiscounted Cash Flow (7)-(13) (14) ($MM)	Discounted Cash Flow @12% (10)*(14) (15) ($MM)	Cumulative Discounted Cash Flow @12% (16) ($MM)	Time (years)
1997	4.678	0.9449	4.420		4.678	-4.678	-4.420	-4.420	1
1998		0.8437	0.000	0.318	2.068	6.681	5.636	1.216	2
1999		0.7533	0.000	0.318	1.269	3.486	2.626	3.842	3
2000		0.6728	0.000	0.318	2.116	6.876	4.624	8.467	4
2001		0.6005	0.000	0.318	2.909	10.044	6.032	14.498	5
2002		0.5362	0.000	0.318	2.710	9.248	4.959	19.457	6
2003		0.4787	0.000	0.318	2.354	7.827	3.747	23.204	7
2004		0.4274	0.000	0.318	2.137	6.958	2.974	26.178	8
2005		0.3816	0.000	0.318	1.928	6.123	2.337	28.515	9
2006		0.3407	0.000	0.318	1.709	5.245	1.787	30.302	10
2007		0.3042	0.000	0.318	1.524	4.504	1.370	31.673	11
2008		0.2716	0.000	0.318	1.249	3.405	0.925	32.598	12
2009		0.2425	0.000	0.318	1.219	3.286	0.797	33.395	13
2010		0.2165	0.000	0.318	1.219	3.286	0.711	34.106	14
2011	0.204	0.1933	0.039	0.318	1.423	3.082	0.596	34.702	15
Total	4.882		4.460	4.452	30.511	75.374	34.702		

Producing Tax Rate (%) = 20.00

Interest Rate (%)	Discounted Cash Flow ($MM)
Value for Example Above = 12.00	34.702
Starting Interest Rate = 10.00	
10.00	38.954
20.00	22.781
30.00	14.455
40.00	9.675
50.00	6.699
60.00	4.728
70.00	3.360
80.00	2.373
90.00	1.640
100.00	1.081
110.00	0.646
120.00	0.302
130.00	0.026
140.00	-0.199
150.00	-0.384
160.00	-0.537
170.00	-0.665
180.00	-0.773
190.00	-0.864
200.00	-0.942
Ending Interest Rate = 200.00	

Payout Time (years) = 1.78
Profit-to-Investment Ratio = 16.44
Present Worth Net Profit ($MM) = 34.702
Present Worth Index = 7.781
Discounted Cash Flow Return on Investment (%) = 131.15

Table 18-8 • Economic Evaluation of Case 2

ECONOMIC EVALUATION RESULTS

	Case-1	Case-2	Case-3	Case-4	Case-5
Capital Investment, $MM	1.853	4.882	0.973	3.484	8.799
Reserves, MMSTBO	1.965	5.138	1.378	3.176	5.105
Project Life	15	15	15	15	15
Payout, Years	2.58	1.78	2.44	2.74	2.28
Discounted Cash Flow Return on Investment, %	69.64	131.15	80.12	87.83	104.84
Present Worth Net Profit, $MM	9.454	34.702	7.013	18.721	35.184
Profit-to-Investment Ratio	16.88	16.44	23.32	13.91	8.74
Present Worth Index	5.66	7.78	8.02	5.90	4.35
Development Costs, $/STBO	0.94	0.95	0.71	1.10	1.72

Table 18-9 • Economic Evaluation Results

The very nature of economic evaluation entails risk taking and uncertainties involving technical, economic, and political conditions. The results of the analysis are subjected to many restrictive assumptions in forecasting recoveries, oil and gas prices, investment and operating costs, and inflation rate. Unforeseen national and world economic and political climates can also severely effect the outcome of the projects.

Figure 18-17 shows the sensitivities of DCFROI and PWNP to the oil price, oil production, investment, and operating costs. The analysis shows that DCFROI is affected more drastically by oil price, oil production, and investment than by the operating costs. PWNP is most sensitive to oil price and oil production. Note that the data points for operating costs and capital investments are nearly coincident in this plot.

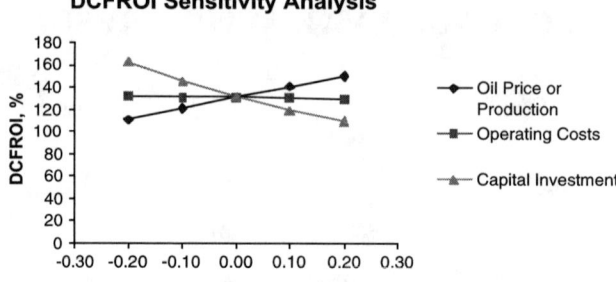

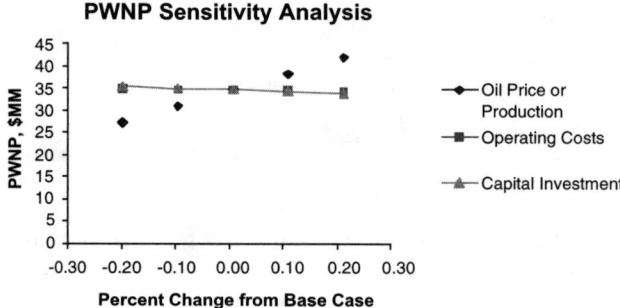

Fig. 18-17 • Sensitivity Analysis of Case 2

REFERENCES

1. Satter, A., and Thakur, G. C: *Integrated Reservoir Management : A Team Approach*, PennWell Books, Tulsa, OK (1994)

Chapter NINETEEN

MINI-SIMULATION EXAMPLES

INTRODUCTION

Historically, reservoir simulators have been used for studying full-field production performance. These studies are costly, requiring highly trained professionals, and are often too time-consuming for operating department environments. Simulators are now available on personal computer platforms to allow access to a single most versatile tool—mini-simulation.

Mini-simulation can play an important role in everyday operations, such as:

- single vertical well performance
- single horizontal well performance
- single well coning performance
- pattern waterflood performance

This chapter will be devoted to appli-

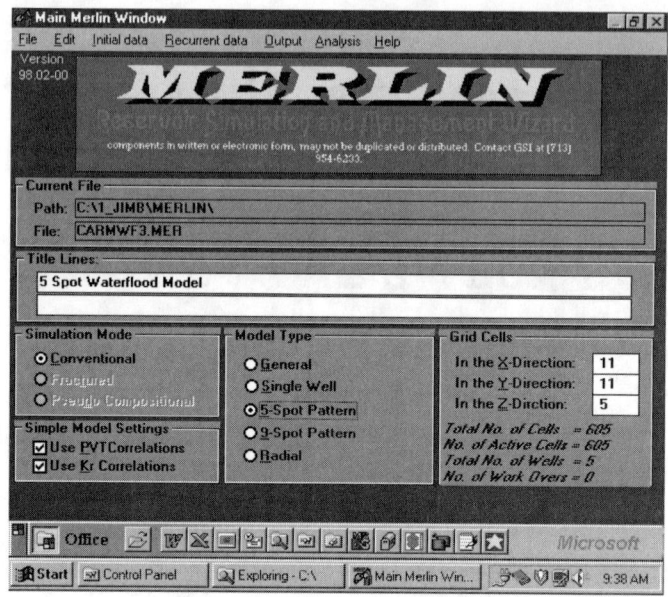

Fig. 19-1 • Merlin Reservoir Simulation and Management Wizard

cations of mini-simulation using a commercial black oil simulator. The influence of various factors on recovery performance will be analyzed. The approach to the sensitivity study will be to build a base case and then to analyze the performance as a specific parameter is varied. Understanding the effects of these factors on recoveries is essential in designing, implementing, monitoring, and evaluating the project. In order to demonstrate the mini-simulations, a reservoir simulation program known as MERLIN (from Gemini Solutions) is used (Fig. 19-1).

For each of these mini-simulations, the same basic rock, fluid, and reservoir description data were used. Fluid data for oil and gas were based upon correlations using basic input reservoir temperature, oil and gas specific gravity, rock compressibility, initial solution GOR, and others:

Reservoir temperature .... 123° F
Rock compressibility 3.0 E-06/psi
Oil density .... 33° API

Gas specific gravity 0.6773

Initial solution GOR . . . . 350 stb/scf

Bubble point . . . . 1,855 psia

Oil viscosity . . . . 1.56 cp

General oil PVT properties as calculated by the Merlin pre-processor are shown in Figure 19-2, and the gas PVT properties are shown in Figure 19-3. Relative permeability data were developed using correlations and basic saturation endpoints as shown below:

Connate water saturation, S_w 0.220

Residual oil saturation to gas, S_{org} 0.010

Residual oil saturation to water, S_{or} 0.200

Critical gas saturation, S_{gc} 0.020

Relative permeability to water, K_{rw} 1.000

Relative permeability to oil in water, K_{row} . . 0.350

Relative permeability to oil, K_{ro} 1.000

Relative permeability to oil in gas, K_{rog} 1.000

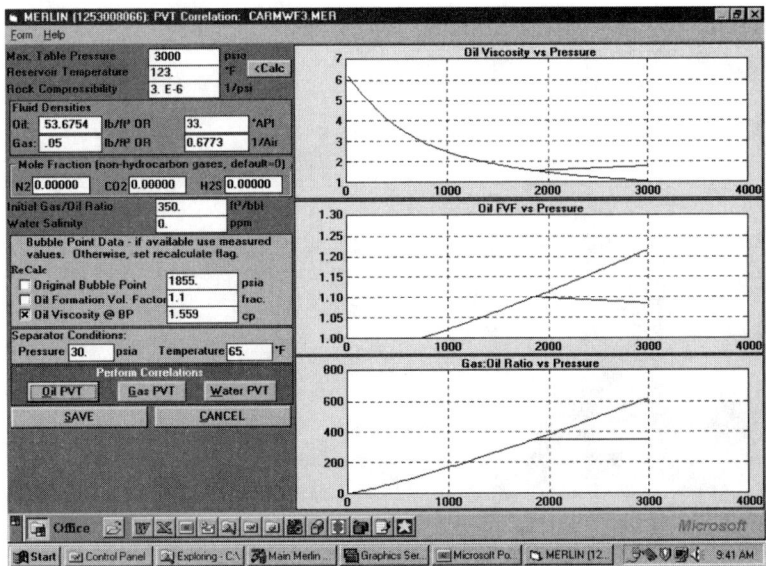

Fig. 19-2 • Merlin PVT Oil Properties

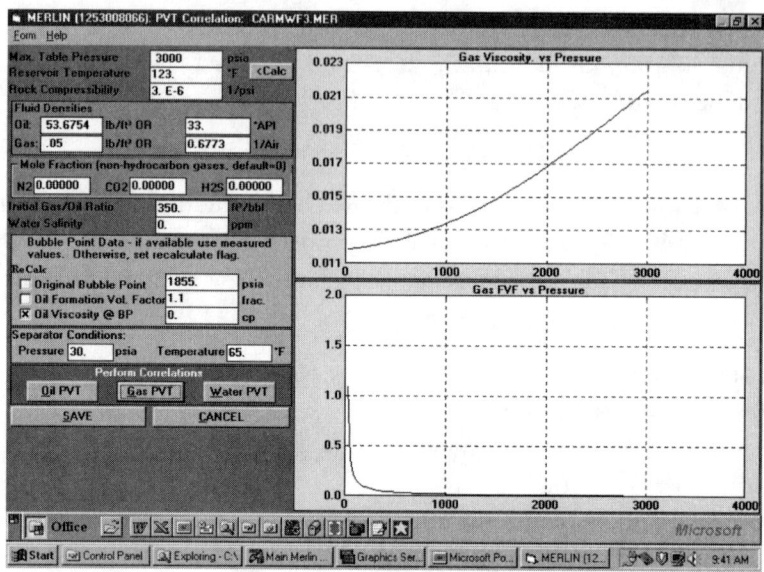

Fig. 19-3 • Merlin PVT Gas Properties

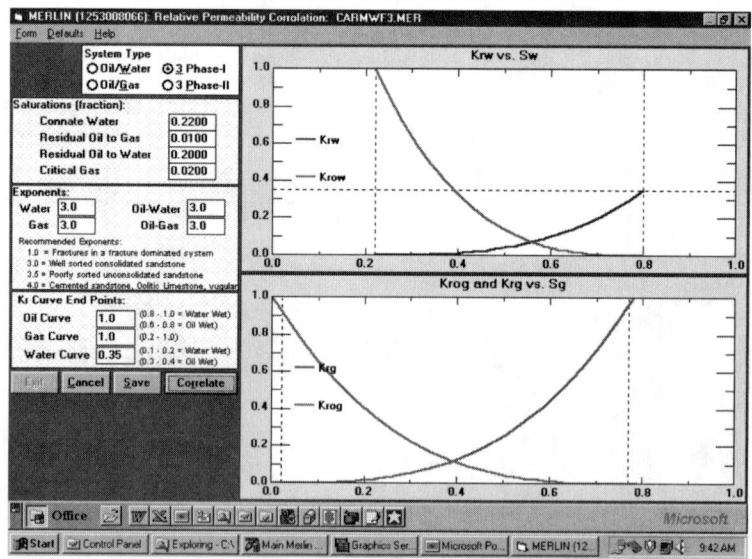

Fig. 19-4 • Merlin Oil-Water and Gas-Oil Permeability Correlation

Figure 19-4 shows the oil-water and gas-oil two-phase relative permeability curves developed by the Merlin simulator.

All of the mini-simulations used the same general reservoir description for a 5-layer model. The changes were in drainage area, simulation grid dimensions, and/or radial vs. Cartesian coordinates. These properties are summarized in Figure 19-5. The vertical permeability (K_z) is set at 0.10 times horizontal permeability (K_x, K_y).

SINGLE VERTICAL WELL

Production performance of a vertical well can be analyzed with a mini-simulation model to show the effects of drainage radius, layering, completion, rock, and fluid properties. This study will be very useful in case of a new discovery with limited data available. A radial model was developed for an oil well in the center of the field, which is completed in all five layers. The model grid was radial and dimensioned 10x5. There was no oil/water contact and the reservoir was initially above the bubble point (no gas cap). This

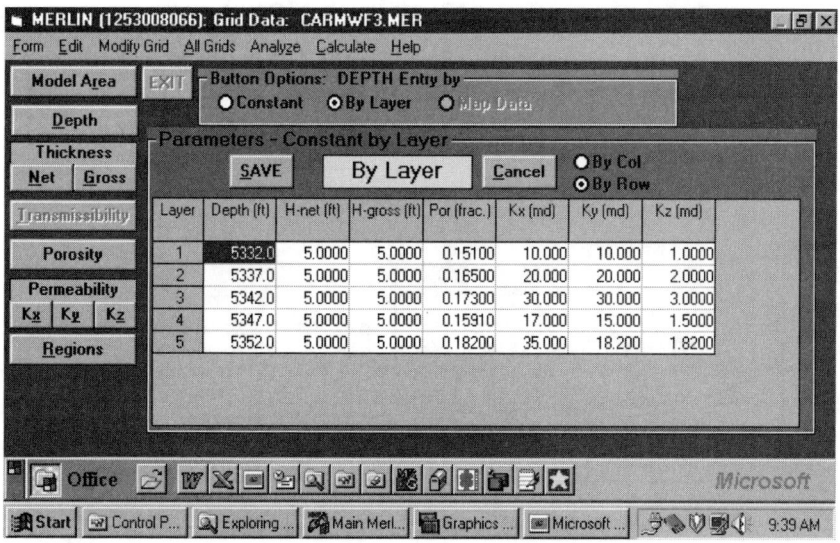

Fig. 19-5 • Merlin Grid Data

example varied only drainage radius from 160 acres to 640 acres. In-place volumes were as follows:

	Case 1	Case 2
Drainage area	160 acres	640 acres
Oil - MMSTB	3.67	14.69
Gas - BCF	1.29	5.14

For both models, the production well was initialized at an oil rate of 1,000 bbls/day. Figure 19-6 plots oil rate in barrels/day. It shows, while both wells can make the initial oil rate, performance for the 160-acre drainage area has a rapid deterioration. This is because the larger drainage area allows higher sustained rates under a normal depletion drive reservoir mechanism. Figure 19-7 plots GOR. In the case with the smaller drainage area, it shows that gas breakthrough is observed at about 1,200 days as a result of the formation of a secondary gas cap. Figure 19-8 plots average reservoir pressure and shows when each case drops below the bubble point. This occurs at

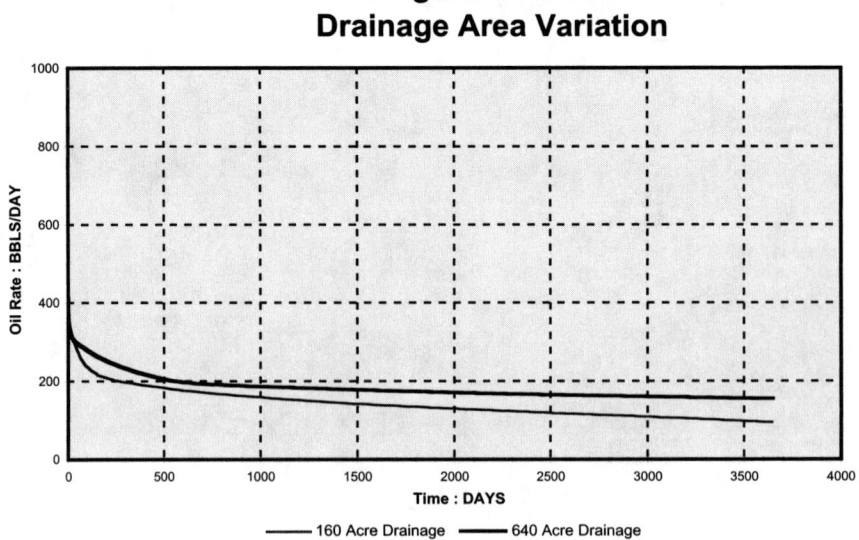

Single Well Model
Drainage Area Variation

Fig. 19-6 • Predicted Vertical Well Oil Production

approximately 147 days for the 160-acre drainage case versus 548 days for 640-acre drainage. These figures show how a well in a larger drainage pattern would perform in contrast to a smaller drainage pattern. As observed, the larger drainage area allows for higher rates, more recovery, and slower decline in overall reservoir pressure.

At the end of a 10-year production period, we have the following comparison:

	160 Acres	640 Acres
Initial reservoir pressure, psia	2,337	2,337
Final reservoir pressure, psia	1,235	1,672
Initial oil rate, b/d	1,000	1,000
Final oil rate, b/d	95	155
Final gas rate, mcf/d	115	49
Cum. oil produced, mstb	521.3	666.5
Cum. gas produced, mmscf	237.0	208.0

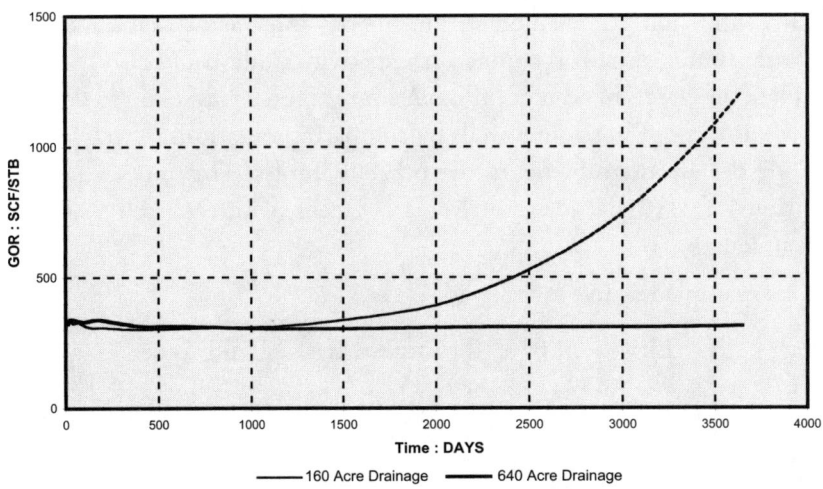

Single Well Model
Drainage Area Variation

Fig. 19-7 • Predicted Vertical Well Gas-Oil Ratio

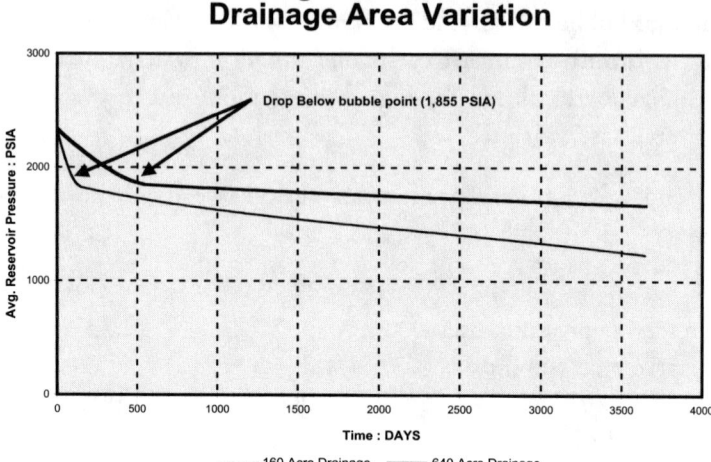

Fig. 19-8 • *Predicted Vertical Well Reservoir Pressure*

SINGLE-LATERAL HORIZONTAL WELL

Potential production performance of a horizontal well can be analyzed to show the effects of well length, zone thickness, vertical-to-horizontal permeability ratio, and the position within the pay interval, as well as distance from an oil/water or gas/oil contact. A sensitivity study could also be made to evaluate the effects of aquifer influx. A simple model was set up to demonstrate the analysis of horizontal wells by comparing variations in well length. All properties are the same as in the other examples. This model has grid dimensions of 21x21x5, for a total of 2,205 cells, with each cell 528 ft by 528 ft in length.

This example has five layers:

Layer	Elevation, ft	Thickness, ft	Fluid Type
1	-5,332	5	Oil
2	-5,337	5	Oil
3	-5,342	5	Oil
4	-5,347	5	Water
5	-5,352	5	Water

There is an oil/water contact at -5,347 at the top of layer 4. The reservoir is above the bubble point, and there is no gas cap or active aquifer. The horizontal well is centered within the grid as shown in Figure 19-9. In-place volumes are as follows:

Oil, mmstb 34.6
Water, mmstb 44.2
Gas, bcf 12.1
Free gas, bcf 0.00
Initial pressure, psia 2,335

The horizontal well was completed in the middle of layer 2 and was assigned an initial rate of 2,000 bbls/day of oil. Horizontal well length was the only variable that was investigated, with three variations in length. Simulation runs were over a 10-year period and are summarized below:

Run	A	B	C
Horizontal length, ft	3,800	2,900	1,850
Final pressure, psia	1,469	1,521	1,587
Final oil, b/d	211	190	162
Final gas, mcf/d	149	116	82
Final water, b/d	386	354	307
Final GOR	707	607	504
Final water cut	0.64	0.66	0.66
Cumulative oil, mstb	1,528.7	1,278.7	998.2
Cumulative gas, mmscf	662.7	517.1	368.3
Cumulative water, mstb	1,952.5	1,736.4	1,440.8

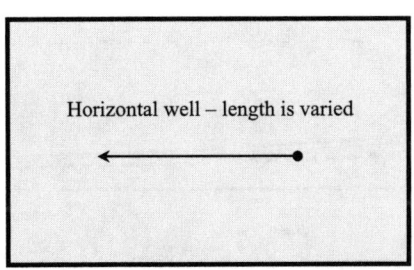

Horizontal well – length is varied

Fig. 19-9 • Horizontal Well Example—Areal view

This group of runs shows how recovery is affected by the length of the horizontal well—all other factors remaining the same. Figure 19-10 plots oil rate vs. time. It clearly shows the longer horizontal well can sustain the initial 2,000 bbl/day oil rate for a longer period of time. As might be expected, cumulative oil production was greater for the well with the horizontal length of 3,800 ft. At 1,528.7 MSTB, it was 53% greater than the oil recovery for the 1,850 ft well length.

Figure 19-11 plots water cut fraction versus time and shows the well with the longer horizontal length has a lower overall water cut. This is due to the overall lower pressure drop attributed to the longer horizontal well length for the same initial oil rate. Water break-through occurs at 50 days for the horizontal length of 1,850 ft, 71 days for a length of 2,900 ft, and 147 days for the length of 3,800 ft. The shorter the horizontal length, the greater the overall pressure drawdown and thus a greater tendency for water to cone.

Figure 19-12 relates GOR vs. time for the three horizontal lengths. As a result of production, overall reservoir pressure is dropped and a secondary gas cap is formed. Gas break-through occurs at 44 days for the shortest well

Horizontal Well Model

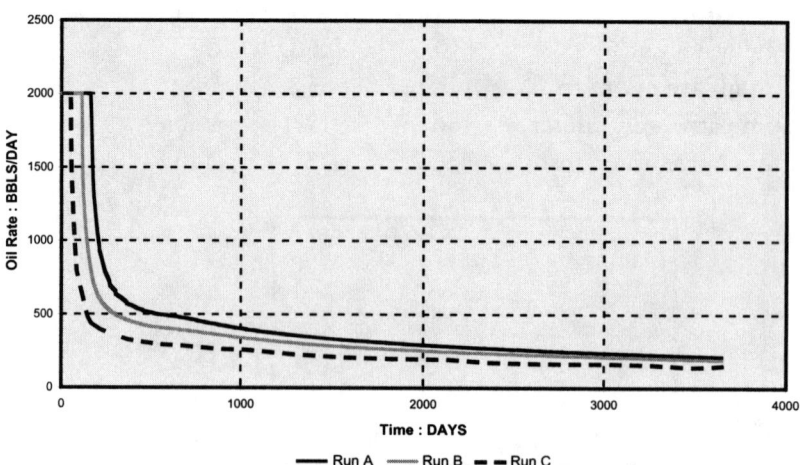

Fig. 19-10 • Predicted Horizontal Well Oil Production Rate

Horizontal Well Model

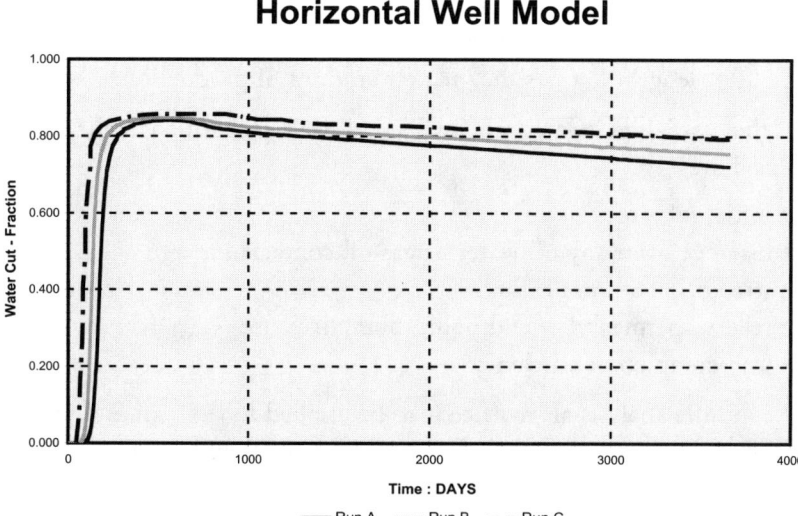

Fig. 19-11 • Predicted Horizontal Well Water Cut

Horizontal Well Model

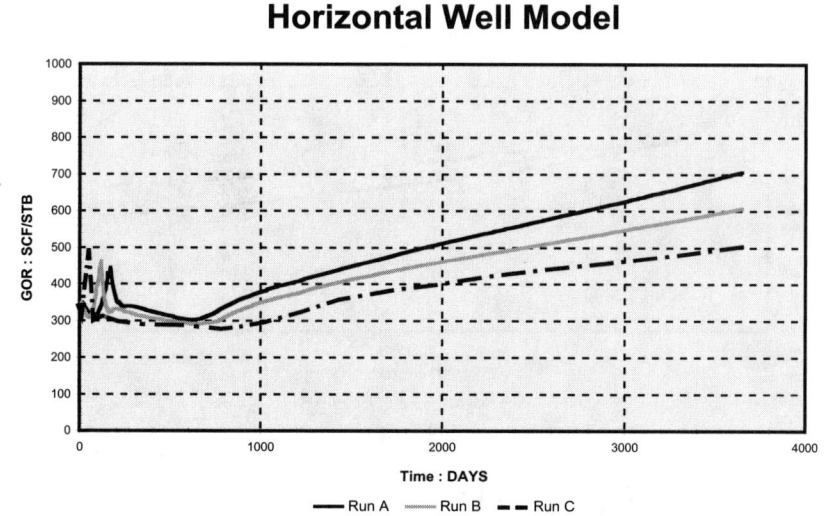

Fig. 19-12 • Predicted Horizontal Well Gas-Oil Ratio

243

length (1,850 ft) to 125 days for the 3,800 ft length. Figure 19-13 relates average reservoir pressure and shows that the greatest drawdown occurs for the 3,800 ft length that has the greatest amount of production.

Other variables related to horizontal well performance that can be evaluated are:

- initial oil rate
- distance from any oil-water or gas-oil contacts
- strength of aquifer influx
- relative permeability end points and curvature
- well completion efficiency

Economic analysis always needs to be applied to determine if the additional oil recovery as a result of additional horizontal well length is worth the incremental drilling cost.

Horizontal Well Model

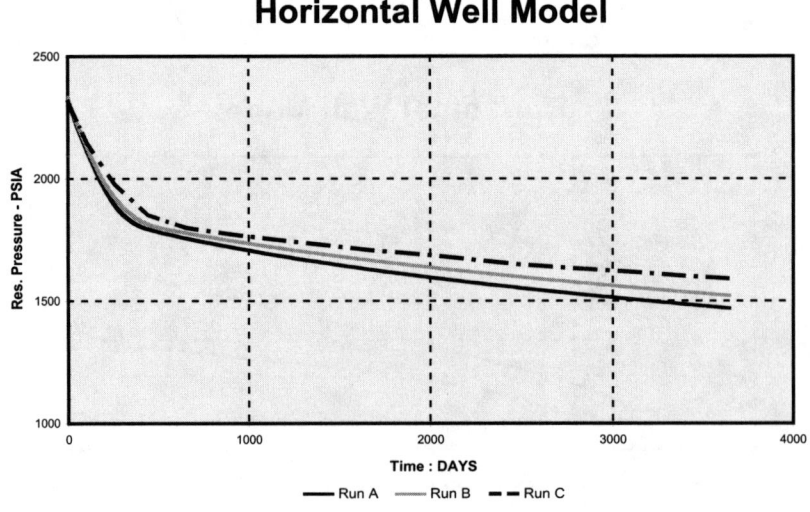

Fig. 19-13 • Predicted Horizontal Well Reservoir Pressure

SINGLE WELL—WATER CONING

Production performance of a well possibly subject to water coning can be analyzed to study the effects of completion interval, vertical to horizontal permeability ratio, production rate, drainage area as well as relative permeability effects. This example uses a single well radial model to evaluate completion interval and tendency to cone water. All rock and fluid properties are the same as in previous examples.

There are five equal layers, each five feet in gross thickness with a net/gross ratio = 1.0. As shown in Figure 19-14, the well is in the center of the radial drainage area and can be completed in any or all of the five layers. There was an oil/water contact located at the interface of layers 3 and 4 that resulted in the model having 10 feet below the oil/water contact and 15 feet above the contact. This mini-simulation looked at the effects of a well completed in the following layers:

Run	Layers
A	1-5
B	1-4
C	1-3

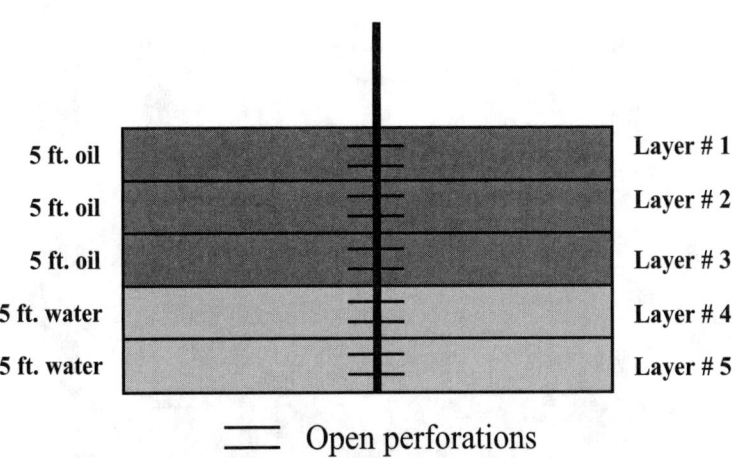

5 ft. oil	Layer # 1
5 ft. oil	Layer # 2
5 ft. oil	Layer # 3
5 ft. water	Layer # 4
5 ft. water	Layer # 5

▬▬ Open perforations

Fig. 19-14 • Producing Well Radial Drainage Area

The radial drainage area was set to be 160 acres with in-place volumes as follows:

Oil, mmstb 2.16
Water, mmstb 2.76
Gas, bcf 0.76
Free Gas, bcf 0.00
Initial Pressure, psia 2,335

Figure 19-15 shows the dimensions of the radial model as 10x5 for a total of 50 grid cells, and also shows how the radial spacing was set up.

The well was set to have an initial oil production rate of 200 bbls/day. Variations in completion interval were investigated. Additional runs could be made to evaluate optimum producing rate for a given completion interval. Figure 19-16 plots the oil rate and shows that a higher oil rate can be sustained with a longer completion interval, but after 10 years of production, the final oil rate is essentially the same for all three cases. A water-cut

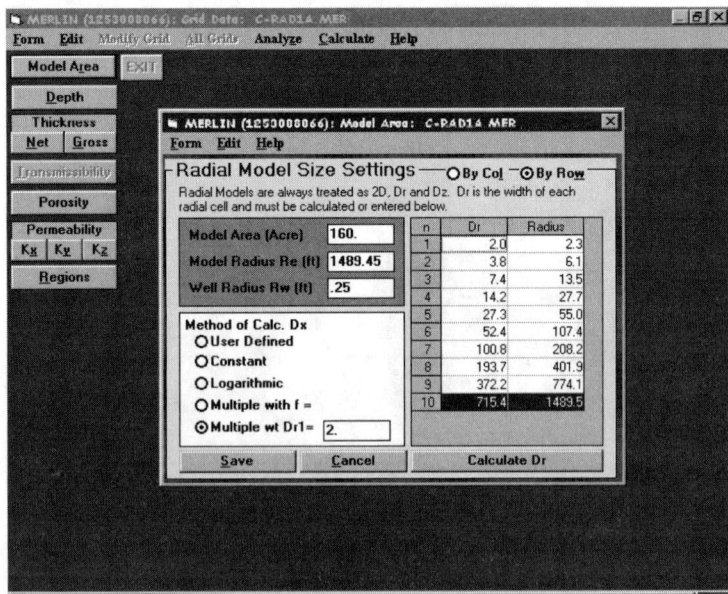

Fig. 19-15 • Radial Model for Coning

plot in Figure 19-17 shows that all cases initially produce at high water cuts, *i.e.*, a significant tendency to cone. The completion interval over all five lay-

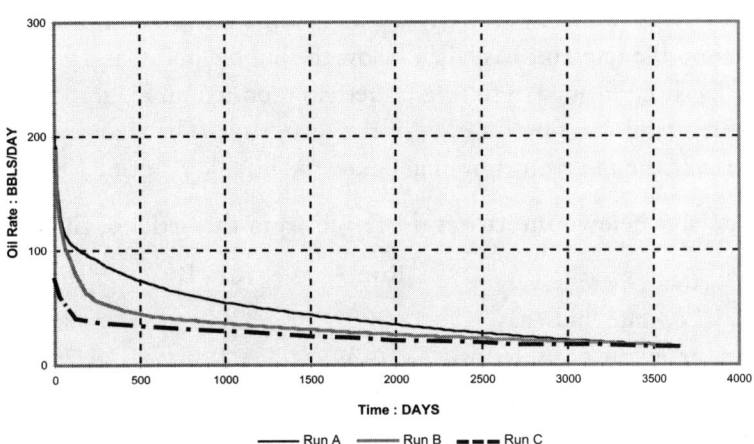

Fig. 19-16 • Coning Well Oil Production

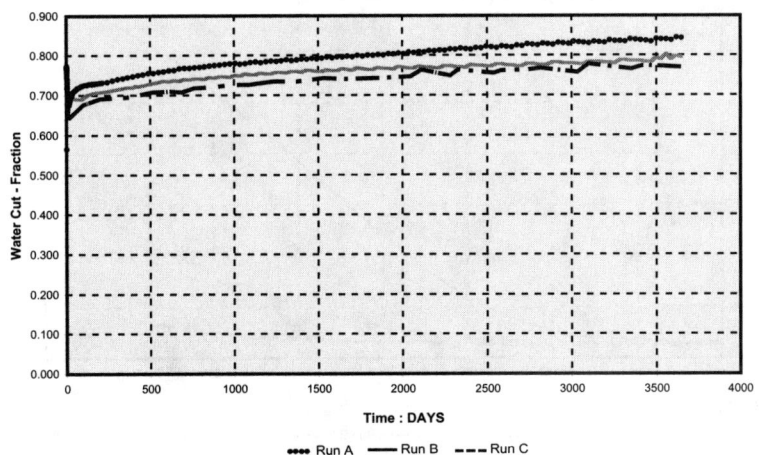

Fig. 19-17 • Coning Well Water Cut

ers (including 2 layers below the oil/water contact) had the highest initial water cut and remained the highest during the 10 year producing period. The lowest water cut was achieved with the completion interval only in the three layers above the oil/water contact. Figure 19-18 is a GOR plot which shows that Run A (completed in all five layers) has a large increase in gas production at approximately 600 days. Due to large amounts of oil and water production, the reservoir has fallen below the bubble point, and a secondary gas cap has been formed, resulting in free gas production. Figure 19-19 plots average reservoir pressure for each of the three runs. The change in slope is indicative of the reservoir dropping below the bubble point of 1,855 psia.

The table below summarizes the results from this series of runs:

	Run A	Run B	Run C
Initial oil rate, bbls/day	200	200	200
Final reservoir pressure, psia	743	1,120	1,794
Final oil rate, bbls/day	16.8	16.8	18.2
Final gas rate, mscf/day	111.3	466	16.1

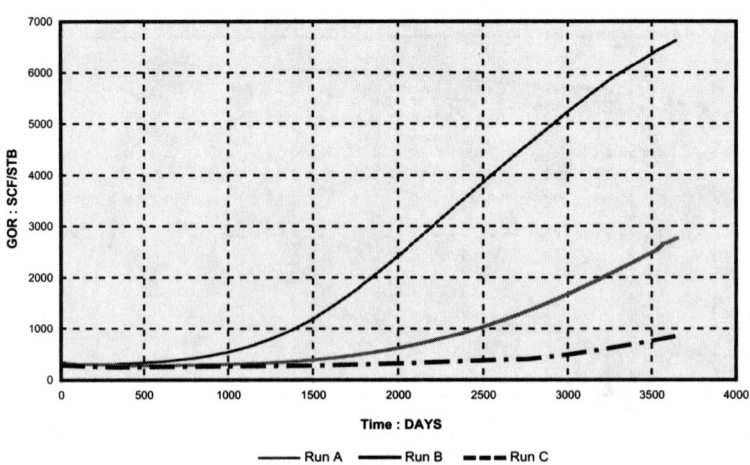

**Single Well Model
Completion Variation**

Fig. 19-18 • Coning Well Gas-Oil Ratio

Final water rate, bbls/day	86.4	62.9	60.6
Cum. oil production, mstb	163.2	125.7	100.3
Cum. water production, mstb	262.5	81.8	37.8
Cum. gas production, mmscf	587.9	375.6	275.7

Other variations can be made using this type of model to evaluate possible water or gas coning. The effects of vertical permeability, as well as the effects of the actual relative permeability curves (end points and curvature), could be studied as a sensitivity. The overall goal would be to optimize recovery with minimal coning. This sort of analysis can be important if production facilities have limited water handling capacity or if water disposal is costly.

5-SPOT WATERFLOODING

An in-depth analysis of waterflood recoveries is influenced by factors such as:

• timing of flood
• layering attributed to depositional environment

Single Well Model
Completion Variation

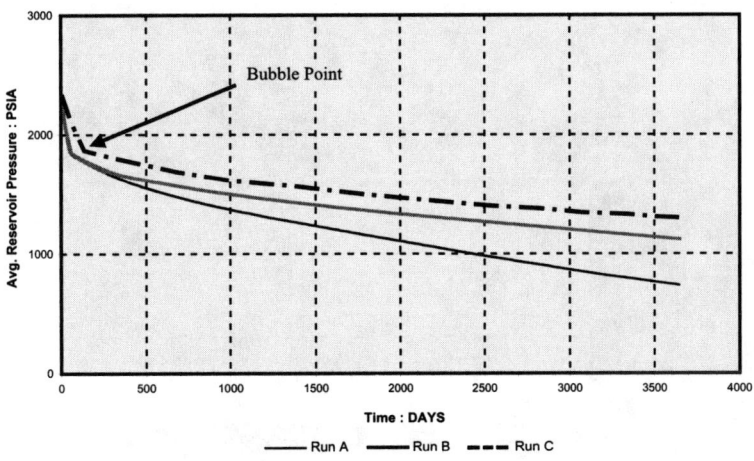

Fig. 19-19 • Average Reservoir Pressure

- vertical permeability variation
- free gas saturation
- cross flow due to vertical permeability variation
- oil gravity
- voidage replacement ratio
- producer/injector ratio
- injection pattern
- increased pattern density

Initial runs were made for a 160-acre 5-spot pattern in an 11x11x5 simulation grid, resulting in a simple model of 605 cells. The injection wells were on the corners and the production well was in the center, as shown in Figure 19-20. This reservoir had no oil/water contact and the pressure was above the bubble point (no initial gas cap) with in-place volumes as follows:

Oil, mmstb 3.67
Water, mmstb 1.12
Gas, bcf 1.29
Free gas, bcf 0.00
Initial pressure, psia 2,335

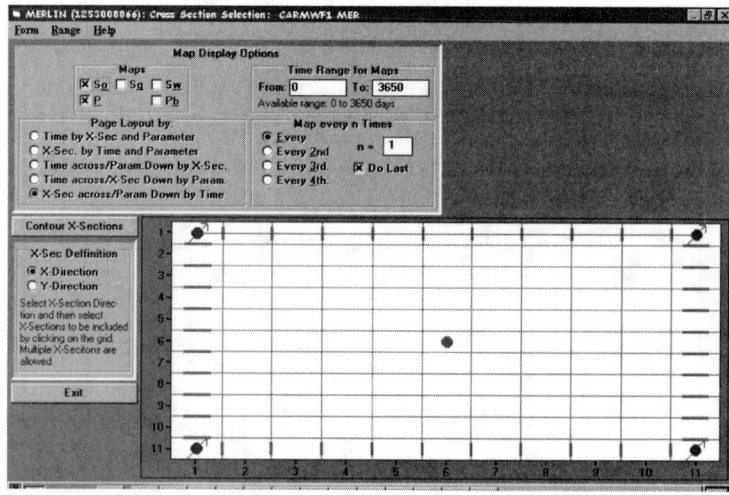

Fig. 19-20 • 5-Spot Waterflood Model

The first run was straight depletion with no pressure maintenance and production from the single producer in the center of the field. The initial oil-producing rate was set at 500 bbls/day. The second run initiated water injection at approximately 4,745 days (13 years), where the reservoir pressure had declined to 717 psia. Water was injected at the four corners in a typical 5-spot operation at a maximum injection pressure of 2,300 psia. Pattern oil recovery was 18% under straight depletion and 45% with water injection. The 45% recovery was achieved after a total of 20,000 days of production and 15,255 days of water injection. Figure 19-21 plots oil rate vs. time. This plot shows after water injection is initiated, approximately 2,000 days are required for fill-up to occur and for oil production to increase.

Figure 19-22 plots produced water rate. There is no water production until 10,585 days or 5,840 days after water injection has started. Figure 19-23 plots average reservoir pressure. Once fill-up has occurred, reservoir pressure increases, since total volume of water injected is greater than the reservoir voidage from production.

One obvious conclusion from this group of runs is that, while a 45% recovery factor from water injection is quite good, the production time

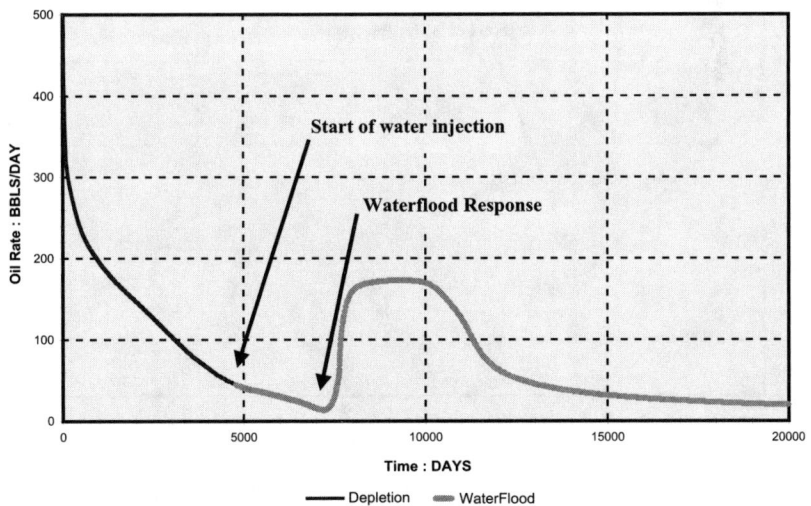

Fig. 19-21 • 5-Spot Waterflood Oil Production Rate

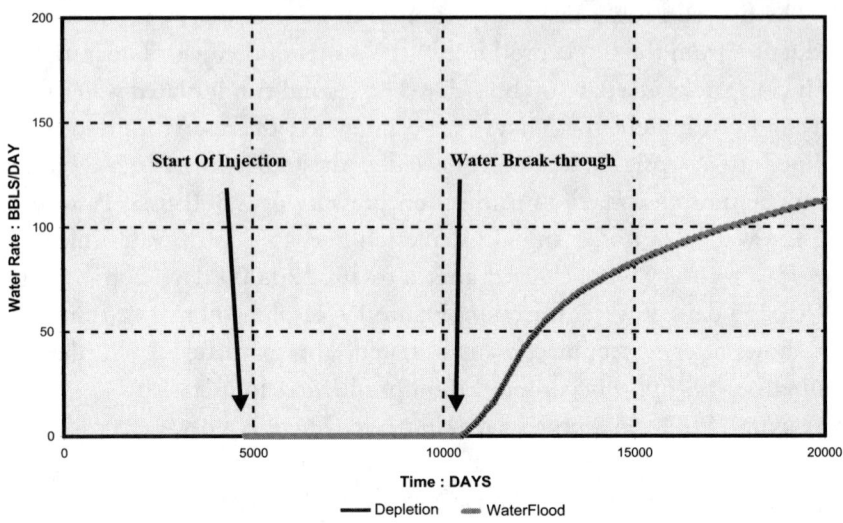

Fig. 19-22 • 5-Spot Waterflood Water Production Rate

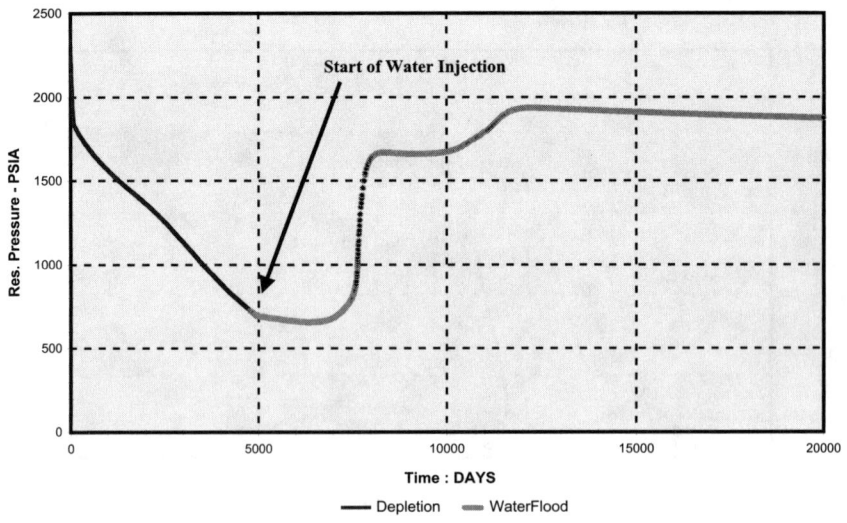

Fig. 19-23 • 5-Spot Waterflood Reservoir Pressure

frame of some 20,000 days is much too long. One single 5-spot pattern is not adequate for the timely sweep of a 160-acre drainage area. Increased pattern density is indicated. With these types of models, additional patterns can be incorporated using increased grid density. For a more in-depth example of a waterflood pattern development, refer to chapters 5-7 of Thakur and Satter.[1]

REFERENCES

1. Thakur, G. C., and Satter, A.: *Integrated Waterflood Asset Management*, PennWell Books, Tulsa, Oklahoma (1998)

index

g

p

r

W